MONITORING ANTAGONISTIC FUNGI DELIBERATELY
RELEASED INTO THE ENVIRONMENT

Developments in Plant Pathology

VOLUME 8

The titles published in this series are listed at the end of this volume.

Monitoring Antagonistic Fungi Deliberately Released into the Environment

Edited by

DAN FUNCK JENSEN

Department of Plant Biology,
Plant Biology Section,
The Royal Veterinary and Agricultural University,
Frederiksberg, Denmark

HANS-BÖRJE JANSSON

Department of Ecology,
Lunds University,
Lund, Sweden

and

ARNE TRONSMO

Department of Biological Sciences,
Agricultural University of Norway,
Ås-NLH, Norway

PARIS

KLUWER ACADEMIC PUBLISHERS
DORDRECHT / BOSTON / LONDON

A C.I.P. Catalogue record for this book is available from the Library of Congress.

ISBN 0-7923-4077-9

Published by Kluwer Academic Publishers,
P.O. Box 17, 3300 AA Dordrecht, The Netherlands.

Kluwer Academic Publishers incorporates
the publishing programmes of
D. Reidel, Martinus Nijhoff, Dr W. Junk and MTP Press.

Sold and distributed in the U.S.A. and Canada
by Kluwer Academic Publishers,
101 Philip Drive, Norwell, MA 02061, U.S.A.

In all other countries, sold and distributed
by Kluwer Academic Publishers Group,
P.O. Box 322, 3300 AH Dordrecht, The Netherlands.

Printed on acid-free paper

Printed in the Netherlands

CONTENTS

GROUP DISCUSSION *IN PLENUM*

PREFACE

This book contains the scientific information presented at the workshop: "Monitoring Antagonistic Fungi deliberately released into the Environment", held at The Royal Veterinary and Agricultural University in Copenhagen, Denmark, on April 1-3, 1995.

Microorganisms, including those genetically modified (GMOs), are increasingly being deliberately released into the environment for different purposes. One major interest is the use of antagonistic organisms for biological control of pests and plant diseases. Both in the development of biocontrol agents and their later risk assessment and product approval, there is a need for methods that facilitate monitoring of such introduced microorganisms. This book particularly addresses the various approaches for monitoring introduced fungi and the current status of techniques which may be relevant to use are described. Such techniques include marker genes and other DNA-based techniques, serology using polyclonals or monoclonals, the use of enzymes or secondary metabolite profiles, "electronic noses" as well as more traditional methods like dilution plating.

The organization of the workshop was done as a joint collaboration between five research groups in Finland, Norway, Sweden and Denmark and the following formed the organizing committee:
Merja Penttilä, Technical Research Centre of Finland (VTT), Finland,
Hans-Börje Jansson, Lunds University, Sweden,
Arne Tronsmo, Agricultural University of Norway,
Dan Funck Jensen, The Royal Veterinary and Agricultural University (RVAU), Denmark.

The local organizers were:
Helge Green, Charlotte Thrane, Mette Lübeck and Peter Stephensen Lübeck, all from RVAU.

In planning the scientific program we were fortunate to have professor James M. Lynch, University of Surrey, Guildford, UK, as the scientific consultant from OECD.

ACKNOWLEDGEMENTS

Thanks are due to the following organizations for supporting the workshop financially:
Nordic Academy for Advanced Study (NorFa) and the
Organisation for Economic Co-operation and Development (OECD).

Dan Funck Jensen,
Chairman of the organizing committee

LIST OF PARTICIPANTS

Sergey Alekseevich Bulat
Lab. of Eukaryote Genetics
Dept. Mol. & Radiation Biophysics,
Petersburg Nuclear Physics Institute,
RAS, Leningrad District, Gatchina,
188350 Russia.
Phone +7 (812) 713 66 25
Fax   +7 (812) 713 71 96
E-mail: bulat@omrb.pnpi.spb.ru

Lars Bødker
Ministry of Agriculture and Fisheries,
Danish Institute of Plant and
Soil Science,
Department of Plant Pathology and Pest
Management, Lottenborgvej 2,
DK-2800 Lyngby
Denmark.
Phone (+45) 45 87 25 10
Fax   (+45) 45 87 22 10

Frances M. Dewey
Department of Plant Sciences,
University of Oxford,
South Parks Road,
Oxford OXI 3RB,
England.
Phone (1865) 27 51 16 or 27 51 05
Fax   (1865) 27 50 74
Email:
dewey@trans.plants.ox.ac.uk

Muriel Dunier
OECD representative,
2, Rue André Pascal
75775 Paris Cedex 16
France.
Phone (33-1) 45 24 95 03
Fax   (33-1) 45 24 18 90

Susanne Elmholt
Danish Institute of Plant and Soil
Science, Research Centre Foulum,
Department of Soil Science,
Box 23, DK-8830 Tjele,
Denmark.
Phone (45) 89 99 18 58
Fax   (45) 89 99 18 69
E-mail: se%pvf%plante2@faulum.min.dk

Helge Green
Department of Plant Biology,
Plant Pathology Section,
The Royal Veterinary and
Agricultural University,
Thorvaldsensvej 40, 8, III.
DK-1871 Frederiksberg C,
Denmark.
Phone (45) 35 28 33 06
Fax   (45) 35 28 33 10
E-mail: hg@kvl.dk

Nina Heiberg
Ullensvang Research Centre,
Division Njös
N-5840 Hermansverk
Norway.
Phone (47) 57 65 36 11
Fax   (47) 57 65 40 53

Linda Hjeljord
Norwegian Crop Research Institute,
Plant Protection,
Fellesbygget
1432 Ås
Norway.
Phone (47) 64 94 94 70
Fax   (47) 64 94 92 26
E-mail:linda.hjeljord@planteforsk.nlh.no

John Hockenhull
Department of Plant Biology,
Plant Pathology Section,
The Royal Veterinary and
Agricultural University,
Thorvaldsensvej 40, 8, III.
DK-1871 Frederiksberg C,
Denmark.
Phone (45) 35 28 33 08
Fax   (45) 35 28 33 10
E-mail: jh@kvl.dk

Lillian Holmer
Department of Forest Mycology
and Pathology,
Swedish University of
Agricultural Sciences,
Box 7026 S-750 07 Uppsala
Sweden.
Phone (46) 18 67 18 07
Fax   (46) 18 30 92 45

Hans-Börje Jansson
Department of Microbial Ecology,
Lunds University,
Ecology Building,
S-223 62 Lund
Sweden.
Phone (46) 46 222 96 14
Fax   (46) 46 222 41 58
E-mail:
hans-borje.jansson@mbioekol.lu.se

Lars Bogø Jensen
Department of Generel Microbiology,
University of Copenhagen,
Sølvgade 83 H,
DK-1307 Copenhagen K,
Denmark.
Phone (45) 35 32 20 53
Fax   (45) 35 32 20 40

Dan Funck Jensen
Department of Plant Biology,
Plant Pathology Section,
The Royal Veterinary and
Agricultural University,
Thorvaldsensvej 40, 8, III.
DK-1871 Frederiksberg C,
Denmark.
Phone (45) 35 28 33 04
Fax   (45) 35 28 33 10
E-mail: dfj@kvl.dk

Rasmus Kjøller
Department of Mycology,
University of Copenhagen,
Øster Farimagsgade 2D,
DK-1353 København K,
Danmark.
Phone (45) 35 32 23 14
Fax   (45) 35 32 23 21

Inge M. B. Knudsen
Department of Plant Biology,
Plant Pathology Section,
The Royal Veterinary and
Agricultural University,
Thorvaldsensvej 40, 8, III,
DK-1871 Frederiksberg C,
Denmark.
Phone (45) 35 28 33 06
Fax   (45) 35 28 33 10

Erland Liljeroth
Department of Plant Breeding Research,
The Swedish University of Agricultural
Sciences, S-26831 Svalöv,
Sweden.
Phone (46) 46 222 96 14
Fax   (46) 46 222 41 58

Mette Lübeck
Department of Plant Biology,
Plant Pathology Section,
The Royal Veterinary and
Agricultural University,
Thorvaldsensvej 40, 8, III,
DK-1871 Frederiksberg C,
Denmark.
Phone (45) 35 28 33 20
Fax (45) 35 28 33 10
E-mail: met@kvl.dk

Peter Stephensen Lübeck
Department of Plant Biology,
Plant Pathology Section,
The Royal Veterinary and
Agricultural University,
Thorvaldsensvej 40, 8, III,
DK-1871 Frederiksberg C,
Denmark.
Phone (45) 35 28 33 20
Fax (45) 35 28 33 10
E-mail: pesl@kvl.dk

J. M. Lynch
School of Biological Sciences,
University of Surrey, Guildford,
Surrey GU2 5XH,
England.
Phone (44) 14 83 25 97 21
Fax (44) 14 83 25 97 28
E-mail: j.lynch@surrey.ac.uk

Anne Mette Madsen
Department of Plant Biology,
Plant Pathology Section,
The Royal Veterinary and
Agricultural University,
Thorvaldsensvej 40, 8, III,
DK-1871 Frederiksberg C,
Denmark.
Phone (45) 35 28 33 20
Fax (45) 35 28 33 10

Morten Miller
Department of Generel Microbiology,
University of Copenhagen,
Sølvgade 83 H,
DK-1307 Copenhagen K,
Denmark.
Phone (45) 35 32 20 54
Fax (45) 35 32 20 40

Søren Molin
Department of Biotchnology,
Technical University of Denmark,
Lundtoftevej 100,
DK-2800 Lyngby,
Denmark.
Phone (45) 45 93 12 22 ext 2513
Fax (45) 42 88 49 22

Preben Nielsen
Department of Ecology and Molecular
Biology, Microbiology Section,
Rolighedsvej 21,
DK-1871 Frederiksberg C,
Denmark.
Phone (45) 35 28 26 27
Fax (45) 35 28 26 24

Lotta Persmark
Department of Microbial Ecology,
Lund University,
Ecology Building,
S-223 62 Lund,
Sweden.
Phone (46) 46 222 37 60
Fax (46) 46 222 41 58

Yvonne Persson
Department of Microbial Ecology,
Lunds Universitet
Ecology Building
S-223 62 Lund
Sweden.
Phone (46) 46 222 37 82
Fax   (46) 46 222 41 58
E-mail:
yvonne.persson@mbioekol.lu.se

Johan Schnürer
Swedish University of Agricultural
Sciences, Department of Microbiology,
Box 7025,
S-750 07 Uppsala,
Sweden.
Phone (46) 18 67 10 00
Fax   (46) 18 67 33 92

Stina Petersson
Swedish University of Agricultural
Sciences, Department of Microbiology,
Box 7025,
S-750 07 Uppsala,
Sweden.
Phone (46) 18 67 33 88
Fax   (46) 18 67 33 92

Robin Sen
Division of General Microbiology,
PL 41 (Mannerheimtie 172),
00014 University of Helsinki,
Finland.
Phone (358) 04 73 51
Fax   (358) 04 73 54 26

Søren Rosendahl
Department of Mycology,
University of Copenhagen,
Øster Farimagsgade 2D,
1353 København K,
Danmark.
Phone (45) 35 32 23 14
Fax   (45) 35 32 23 21

Jan Stenlid
Department of Forest Mycology
and Pathology, Swedish University of
Agricultural Sciences,
Box 7026,
S-750 07 Uppsala
Sweden.
Phone (46) 18 67 18 07
Fax   (46) 18 30 92 45

Lone Rossen
Biotechnological Institute,
Lundtoftevej 100,
Building 227,
DK-2800 Lyngby,
Denmark.
Phone (45) 45 87 66 99
Fax   (45) 45 93 28 88

Bengt Söderström
Department of Microbial Ecology,
Lunds University,
Ecology Building,
S-223 62 Lund,
Sweden.
Phone (46) 46 222 86 17
Fax   (46) 46 222 41 58

Chris R. Thornton
Department of Plant Sciences,
Cambridge University,
Downing Site,
Downing Street,
Cambridge CB 2 3 EA,
England.
Phone (44) 1223 33 02 29
Fax (44) 1223 33 39 53
E-mail: crt@cam.ac.uk

Charlotte Thrane
Department of Plant Biology,
Plant Pathology Section,
The Royal Veterinary and
Agricultural University,
Thorvaldsensvej 40, 8, III,
DK-1871 Frederiksberg C,
Denmark.
Phone (45) 35 28 33 04
Fax (45) 35 28 33 10

Ulf Thrane
Department of Biotchnology,
Technical University of Denmark,
Building 221,
DK-2800 Lyngby,
Denmark.
Phone (45) 45 93 30 66
Fax (45) 45 88 49 22
E-mail: ut@ibt.dtu.dk

Sari Sinikka Timonen
Department of Biosciences,
Devision of General Microbiology,
P.O.Box 41,
00014 University of Helsinki,
Finland.
Phone (358) 04 73 54 25
Fax (358) 04 73 54 26

Arne Tronsmo
Department of Biological Sciences,
Agricultural University of Norway,
P.O. Box 40,
N-1432, ÅS-NLH,
Norway.
Phone (47) 64 94 75 00
Fax (47) 64 94 77 50
E-mail: arne.tronsmo@post.nlh.no

Valerie Vasseur
Laboratoire da Microbiologie
et Biochimie,
ESMISAB,
Technopôle Brest-Iroise,
29280 Plouzane,
France.
Phone (33) 98 05 61 26
Fax (33) 98 05 61 01
E-mail: vasseur@univ-brest.fr

Susanne Vestergaard
Department of Ecology and Molecular
Biology,
Zoology Section,
The Royal Veterinary and
Agricultural University,
Bülowsvej 13,
1870 Frederiksberg C,
Denmark.
Phone (45) 35 28 26 84
Fax (45) 35 28 26 70
E-mail:
susanne.vestergaard@ecol.kvl.dk

Helena Åkesson
Department of Microbial Ecology,
Lund University,
Ecology Building,
S-223 62 Lund,
Sweden.
Phone (46) 46 222 37 60
Fax (46) 46 222 41 58

MONITORING OF ANTAGONISTIC FUNGI. PERSPECTIVES, NEEDS AND LEGISLATION

J.M. LYNCH
School of Biological Sciences
University of Surrey, Guildford, GU2 5XH, UK

1. Introduction

The opportunities to harness the antagonistic activities of fungi by deliberately releasing them into the environment is of great interest throughout the world and one which has accelerated with the advancing knowledge in the field of molecular biology. Many national and international groups have taken interest in this field. Particularly the Organisation for Economic Co-operation and Development (OECD), through its Directorate for Food, Agriculture and Fisheries, has a co-operative research programme on Biological Resource Management for Sustainable Agricultural Systems, which has just had funding approved to run for a further five years from 1995-1999. Of the four themes, the first is on Safe Exploitation of Micro-organisms in Plant/Soil Systems. The topics covered within the theme are:

- Methods for molecular ecology
- Identification of physiochemical aspects of the soil environment which regulate microbial function
- Risk analysis and toxicology of the use of micro-organisms
- Production and delivery of microbial inocula
- Reduction of the load of chemical pesticides, fertilisers and organic wastes on the soil ecosystem
- Assessment of biodiversity in plant/soil systems

The programme funds fellowships and workshops. It is obvious that the monitoring of antagonistic fungi is very relevant to most of the topics.

OECD has also been very active in provision of a forum for the debate on the use of genetic engineering techniques and on the release of organisms into the environment.

1

D. F. Jensen et al. (eds.), Monitoring Antagonistic Fungi Deliberately Released into the Environment, 1–9.
© *1996 Kluwer Academic Publishers. Printed in the Netherlands.*

With particular inputs from the Directorates of Science and Technology and of Environment, it first produced a booklet on laboratory safety issues concerning the use of recombinant DNA (OECD, 1986), and this was followed with a booklet on assessing risk following the release of genetically modified plants and microorganisms (OECD, 1992). Most recently a volume has been produced which concerned the scale-up of microorganisms as biofertilisers (OECD, 1995). None of these reports have any legal standing, indeed OECD only ever acts in an advisory role to its member countries. Nevertheless it seems that most countries, including the European Union, have accepted the advice contained within the reports and incorporated them into national regulations. This is not surprising because the reports were drafted by groups of national experts in biotechnology.

Inevitably, in producing the reports the incomplete state of knowledge was identified. This is where the Co-operative Research Programme has had an important role to play. From the workshop element of the programme, published volumes which have included the issues on the mathematical interpretation and prediction of release of organisms into the environment (Bazin and Lynch, 1993), the release of organisms for the biological control of pests and diseases (Hokkanen and Lynch, 1995) and the development of soil inoculants (Elliott and Lynch, 1995). Most recently Lynch and Elliott (1996) have addressed the question of the need for bioindicators as the critical aspect and need for monitoring the environment for both indigenous and introduced organisms and their associated biochemical activities.

In the following brief survey, a few pertinent issues from personal research will be outlined to illustrate some of the complications and opportunities in monitoring antagonistic fungi in soil.

2. Fallacies of Counting Propagule Numbers

Whereas bacteria exist in soil as distinct cells, fungi occur as mycelia and spores. In pure culture, the development of bacterial biomass can usually be monitored satisfactorily by plating aliquots of the culture onto an agar medium and measuring the number of colony-forming units that develop. For fungi, there is no such relationship because, whereas each spore usually gives rise to a colony, only a proportion of the mycelium will give rise to colonies. Inevitably, the situation becomes even more complicated when the microorganisms are in natural environments. Among these complications are:

- determining the proportion of organisms which can actually be cultured on the isolation medium
- separating living, senescent and dead cells
- measuring biomass in solid and opaque substrates, such as soil
- difficulty in distinguishing between fungal and bacterial biomass

It can be expected that the antagonistic action of fungi will be related to the fungal biomass. In a series of experiments using *Trichoderma harzianum* to control the damping-off of lettuce (*Letuca sativa*) caused by *Pythium ultimum*, a variety of fungal biomass markers (ATP, chitin, ergosterol, esterase) were evaluated and compared with viable propagule measurements in the determination of the antagonist/pathogen populations (Lumsden *et al*, 1990). The *Trichoderma* populations in the potting mix increased until four days after inoculation, when it stabilised, whereas the *Pythium* populations declined to almost zero within four days, indicating successful biocontrol of the pathogen by the antagonist. This was also evident from the lack of disease symptoms in the plant when the antagonist was added with the pathogen. Generally, the biomass markers showed an opposite trend to the viable propagule numbers, with the biomass declining quite rapidly. Indeed, there appeared to be an inverse relationship of numbers to biomass. This led to the conclusion that biocontrol activity is primarily linked to only a transient increase in antagonist biomass, but resulting in a sustained increase in active propagules of the antagonist.

3. Chemical Markers

With the limitations of propagule numbers in investigating the ecophysiological functions of soil fungi, a variety of options are available to use chemical markers as determinants of fungal biomass. Probably the most extensively used is chitin, which is contained in all fungal cell walls except the oomycetes. Chitin is measured as N-acetyl glucosamine and the basic method is due to Ride and Drysdale (1972), with subsequent modifications, especially suitable for mycorrhizal fungi by Whipps (1987). This method is of course particularly suitable for looking at antagonistic interactions with the chitin-free pathogen *Pythium* (Lumsden *et al*, 1990).

The fungal sterol, ergosterol, is highly-specific for fungi, but there can be great variation between replicate samples, even when an internal standard is used, possibly because there is no general pattern for regulation of this sterol. Matcham *et al* (1984) found it particularly useful for the basidiomycete *Agaricus bisporus*, while West *et al* (1987) deployed it to monitor changes in fungal populations in soils. In common with other chemical marker methods, time is necessary to standardise the technique in any particular laboratory.

Adenosine triphosphate (ATP) (Eiland, 1985) and fluorescein diacetate (FDA) esterase (Swisher & Carroll, 1980) are not specific for fungi but, depending on soil type and conditions, can be useful indicators of total biomass in soil. It should be considered that the biomass in the bulk soil is dominated by fungi, whereas in the rhizosphere the fungal/bacterial contributions are more equable (Lynch & Hobbie, 1988). Even though the FDA method has been used by many investigators (eg, Schurner & Rosswall, 1982; Soderstrom, 1979; Swisher & Carroll, 1980), it is really an activity measurement rather than a biomass marker. It relies on the uptake of the substrate and release of the product being the same for mycelium of all ages, and the amount of esterase present at all growth stages being constant. These assumptions may not be valid, and thus the assay is probably only reliable under certain well-defined conditions. As FDA is an activity measure, the extra complexity compared with the measurement of respiration probably will not normally make it a good general indicator of activity.

For ease, reliability and specificity of measurement, it would seem that chitin should be considered as a useful indicator of biomass, although with development ergosterol can also be useful. The most extensively used total biomass measure is the fumigation-respiration method of Jenkinson & Powlson (1976) and as there is usually a good correlation with ATP determinations, this is a very suitable biomass indicator.

4. Mycelial Fragments: A New Approach

It is clear from the foregoing that neither propagule counts nor chemical markers are ideal for the determination of fungi in soil. The soil fungal biomass is dominated by mycelia. It seemed that if the mycelia could be fragmented in a Waring blender, as had been achieved for *Penicillum notatum* and *P chrysogenium* (Savage & Vander Brook, 1946) with little effect on growth or fermentative capacity, then the resulting units plated onto solid medium might give rise to colonies related to the fungal biomass present. The residual problem with relatively fast-growing fungi is that they rapidly fill the medium surface with confluent growth. However the paramorphogenic chemical, sodium deoxycholate, has previously been incorporated into solid media to enable colonies to be more readily enumerated because it reduces colony extension rates (Skone & Dixon, 1981).

The use of blending coupled to colony suppression was deployed to investigate the interaction *in vitro* between the antagonist *Trichoderma harzianum* with the pathogen *Fusarium culmorum* (Cheetham *et al*, 1995). The fungi could be distinguished and enumerated on the basis of their colony characteristics, principally the production of pigments. A linear relationship was established between the number fungal colonies produced and the biomass dry weight of each fungus in pure culture, and therefore this facilitated the study of the interaction in mixed culture. The method has yet to be applied to soil systems, but it would seem to offer much promise.

5. Immunology: Problems of Substrate-Dependence and Cross-Reactivity

Potentially, immunology would appear to offer much scope and specificity for the determination of fungi in soil, probably utilising enzyme-linked immunoassay (ELISA) for quantitative estimation. Polyclonal antibodies are usually much less specific and more cross-reactive than monoclonals, but both systems are dependent on the antigen being present in the fungus when growing on all the substrates it encounters and under all the growth conditions. The protein profile of *Trichoderma harzianum* was investigated on glucose and ground straw as carbon substrates. There was a very different pattern on each substrate (Carter & Lynch, 1991), but one suitable marker antigen was used to raise polyclonal antibodies. Although the antisera proved to be of value in the determination of the antagonist biomass, there was some degree of reaction to other strains of *Trichoderma*, *Gliocladium virens* and *Phomopsis sclerotiodes*, but not to a range of other soil fungi. Using monoclonal antibodies to other less-specific antigens, Thornton (1996) has been able to develop assays for the quantification of both *T. harzianum* and *Rhizoctonia solani* in soil. Monoclonal antibodies are less reactive and more expensive to produce than polyclonals, but it is clear that they offer considerably more promise in investigating antagonistic interactions in soil.

6. The Role of Molecular Biology

With the rapidly advancing state of knowledge of molecular biology, a range of probes are now available or can be constructed to follow DNA and RNA in terrestrial microbial populations (Carter & Lynch, 1993). However it must be emphasised that these only give information on the genetic potential of the soil microbial community, and other methods are required to determine whether this potential is being expressed. An advantage is that the probes can be deployed on the total nucleic acid extracted from soil, or on *in situ* or extracted cells. To amplify gene sequences from soil, the polymerase chain reaction can be used (Pillai *et al*, 1991).

In monitoring antagonistic actions at the molecular level, it is important to have genes encoding the antagonism cloned. For antibiotics from fungi, little is known about the genes involved. However there has been some study of chitinase, one of the lytic enzymes thought to be involved in mycoparasitism. An exochitinase (27.5/28 kD) from *Trichoderma harzianum* T198 was purified and sequenced, and a unique 375 base pair fragment found (E.E. Deane, J.F. Peberdy, J.M. Whipps and J.M. Lynch, unpublished). This sequence was found from the EMBL data bank to have 100% homology with the equivalent chitinase amino acid sequence in *Aphanocladium album*, 78% with *T. harzianum* (Chil) and less than 50% with other bacterial sequences. Thus there could be some value in using this partial chitinase clone to probe the population dynamics of the gene in the specific *Trichoderma* strain soil, accepting that there could be some mixture of the signals from other strains. In reality however, the chitinase enzymes are a complex group and a variety of genes code for them. At this stage, we have not correlated the partially cloned gene with lytic or even overall biocontrol activity. There would therefore appear to be limited value in using the DNA probe to investigate antagonism *per se*. Elsewhere in this volume, several authors discuss the DNA probing of antagonistic fungi where there has been a greater realisation of the potential value.

With the incentives to harness the technique of molecular biology, the more classical biochemical methods are frequently ignored. For example, on the issue of chitinase, techniques are well-established which allow the assay of the various forms of the enzyme in soil (Naseby & Lynch, 1996). It is therefore possible to carry out sensible dynamic studies of gene expression by monitoring the enzymes as gene products. Ultimately this is of more direct relevance than the *potential* expression which might be studied with nucleic acid probes.

7. Regulations and Legislation

The use of antagonistic fungi as biocontrol agents would appear to offer considerable opportunity to reduce chemical load on the environment. The agrochemical industry is well set-up to deal with registration of chemicals under the various national and international directives. However the registration of biologicals has presented some complications as to the procedures which need to be used. Clearly, if a biological agent is effective by producing an active substance, it is right and proper that the substance should be evaluated in the same light as a chemical pesticide (Klingauf, 1995). It must be recognised of course that there are many natural products with toxicological properties. The most important requirements for the dossier to be submitted for inclusion is covered under European Union Council Directive of 15 July 1991 (91/414/EEC) concerning the marketing of plant protection products. In Annex 1, part B, microorganisms and viruses, the following need attention:

- Identity of organism
- Biological properties of organism
- Analytical methods
- Toxicology, pathogenicity and infectivity
- Residues in or on treated products, food and feed
- Fate and behaviour in the environment
- Ecotoxicology

Obviously gathering all this information is costly and raises the question as to whether small start-up biotechnological companies would have the necessary resources to finance the registration procedures.

The above considerations for registration assume that the microorganism is not exotic to the country where it is to be evaluated in the field or marketed. In this respect the European Union would be seen as one country, for example, a product isolated and developed in Denmark could be marketed in Great Britain without further complications. However to go outside of the European Union would require regulatory quarantine clearance from the country in which it is to be used. The particular concern on the introduction of exotics has been addressed in a previous OECD Workshop

(Hokkanen & Lynch, 1995). Essentially DNA from an organism in a foreign environment is potentially being introduced to a country.

There has been even greater concern on genetically-modified biocontrol agents being introduced into the environment, irrespective of the country in which the parent wild-type organisms were isolated from. Many of the regulatory procedures in the European Union and other countries have been stimulated by the reports of OECD (1986, 1992, 1995). The critical aspect of regulations is that a release needs to be monitored under the following heads:

- Microorganism profile
  - including genetic construction, ecology and toxicology
- Crop profile
  - including associated insects and wildlife
- Environment profile
  - including soil, meteorology and location
- Introduction profile
  - including production and delivery

Much national and international funding, including that of the OECD Programme, has been focused at deriving and evaluating existing and novel methods to facilitate this monitoring. With the monitoring, the sensible strategy for deployment of exotic or genetically-modified biocontrol agents seems to be (Lynch, 1992):

- Establishment of baseline ecology and diversity
- Determination of spread, survival, benefits and hazards
- Risk analysis
- Production release

8. Conclusions: Needs

Much useful scientific information has been collected on soil antagonistic fungi in recent years. It is perhaps therefore disappointing that there is only one commercial agent currently available. This is *Gliocladium virens* G1-21, registered with the US Environmental Protection Agency by W.R. Grace & Co. Conn. (Lumsden & Walter, 1995). However a range of *Trichoderma* strains have been sold in the past and some are showing considerable potential at present. Perhaps as a scientific community, we should be focusing on the *Gliodladum/Trichoderma* group and ensure there is some significant commercial development. This in itself would provide more stimulus for scientific research. This strategy provided successful in the development of insect control agents which was focused on *Bacillus thuringiensis*, and the bacterium now accounts for about 1.5% of the world crop protection sales market and this is increasing (Cannon, 1995). However there is always a place for quality generic

8

science in any field and hopefully the remaining chapters in this volume will stimulate more focused scientific research towards the successful exploitation of biocontrol for the benefit of environmental health.

9. References

Bazin, M.J. and Lynch, J.M. (eds.) (1993) *Environmental Gene Release: Models, experiments and risk assessment*, Chapman & Hall, London.

Cannon, R.J.C. (1995) *Bacillus thuringiensis* in pest control, in H.M.T. Hokkanen and J.M. Lynch (eds.), *Biological Control: Benefits and Risks*, Cambridge University Press, Cambridge, pp. 190-200.

Carter, J.C. and Lynch, J.M. (1991) Substrate-dependent variation in the protein profile and antigens of *Trichoderma harzianum*, in *Enzyme & Microbial Technol.*, **13**, 557-543.

Carter, J.C. and Lynch, J.M. (1993) Introduction of new immunological and molecular techniques for microbial population and community dynamic studies in soil, in J-M. Bollag and G. Stozky (eds.), *Soil Biochemistry, vol. 8*, Marcel Dekker, New York, pp. 249-272.

Cheetham, J.L., Bazin, M.J., Markham, P. and Lynch, J.M. (1995) A method utilising mycelial fragments to estimate relative biomass densities of fungal species in mixed culture, *J. Microbiol Methods*, **21**, 113-122.

Eiland, F. (1985) Determination of adenosine triphosphate (ATP) and adenylate charge (AEC) in soil and use of adenosine nucleotides, as measures of soil microbial mass and activity, *Report No. 51777*, Statens Planteaulsfarsog, Tidsskrift for Planteauls Specialserie, Copenhagen.

Elliott, L.F. and Lynch, J.M. (1995) The international workshop on establishment of microbial inocula in soils: Cooperative research project on biological resource management of the OECD, *American J. of Alternative Agriculture*, **10**, 50-73.

Hokkanen, H.M.T. and Lynch, J.M. (1995) *Biological control: Benefits and Risks*, Cambridge University Press, Cambridge.

Jenkinson, D.S. and Powlson, D.S. (1976) The effects of biocidal treatments on metabolism in soil: A method for measuring soil biomass, *Soil Biol. Biochem.*, **11**, 193-194.

Klingauf, F.A.J. (1995) Registration requirements of biological control agents in Germany and in the European Union, in H.M.T. Hokkanen and J.M. Lynch (eds.), *Biological Control: Benefits and Risks*, Cambridge University Press, Cambridge, pp. 283-290.

Lumsden, R.D., Carter, J.P., Whipps, J.M. and Lynch, J.M. (1990) Comparison of biomass and viable propagule measurements in the antagonism of *Trichoderma harzianum* against *Pythium ultimum*, Soil Biol. Biochem., **22**, 187-194.

Lumsden, R.D. and Walter, J.F. (1995) Development of the biocontrol fungus *Gliocladium virens*: risk assessment and approval for horticultural use, in H.M.T. Hokkanen and J.M. Lynch (eds.), *Biological Control: Benefits and Risks*, Cambridge University Press, Cambridge, pp. 263-269.

Lynch, J.M. (1992) Environmental implications of the release of biocontrol agents, in E.C. Tjamos, G.C. Papavizas and R.J. Cook (eds.) B*iological Control of Plant Diseases: Progress and Challenges for the Future*, Plenum Press, New York, pp. 389-397.

Lynch, J.M. and Elliott, L.F. (1996) Bioindicators: Perspectives and Potential Value for Landusers, Researchers and Policy Makers, in C.E. Pankhurst, B.M. Doube and V.V.S.R. Gupta (eds.), *Bioindicators of Soil Health*, CABI, Wallingford, Oxon, in press.

Lynch, J.M. and Hobbie, J.E. (1988) *Microorganisms in Action: Concepts and Applications* in *Microbial Ecology*, Blackwell Scientific, Oxford.

Matcham, S.E., Jordan, B.R. and Wood, D.A. (1984) Methods for assessment of fungal growth on solid substrate, in J.M. Grainger and J.M. Lynch (eds.), *Microbial Methods for Environmental Biotechnology*, Academic Press, London, pp. 5-18.

Naseby, D.C. and Lynch, J.M. (1996) Soil enzymes as indicators of perturbations caused by inoculation of a genetically modifed strain of *Pseudomonas fluorescens* SBW25 on wheat seed, Soil Biol. Biochem., submitted.

OECD (1986) *Recombinant DNA Safety Considerations*, OECD, Paris.

OECD (1992) *Safety Considerations for Biotechnology*, OECD, Paris.

OECD (1995) *Safety Considerations for Biotechnology: Scale-up of Crop Plants*, OECD, Paris.

Pillai, S.D., Josephson, R.L., Bailey, R.L., Gerba, C.P. and Pepper, I.L. (1991) Rapid method for processing soil samples for polymerase chain reaction amplification of specific gene sequences, *Appl. Environ. Microbiol.*, **56**, 782-787.

Ride, J.P. and Drysdale, R.B. (1972) A rapid method for the chemical estimation of filamentous fungi in plant tissue, *Physiological Plant Pathology*, **2**, 7-15.

Savage, G.M. and Vander Brack, M.J. (1946) The fragmentation of the mycelium of *Penicillium notatum* and *Penicillium chrysogenum* by a high-speed blender and the evaluation of the blended seed, *J. Bacteriol.*, **52**, 385-391.

Schumer, J. and Rosswall, T. (1982) Fluorescein diacetate hydrolysis as a measure of total microbial activity in soil and litter, *Appl. and Environ. Microbiol.*, **43**, 1256-1261.

Skone, E.J. and Dixon, P.A. (1981) The effect of paramorphogens on growth leinetics of *Ceratocystis adiposa*, *Botrytis fabae* and *Asperigillus chevalieri*, *Microbios*, **32**, 189-202.

Söderstrom, B.E. (1979) Some problems in assessing the fluorescien diacetate-active fungal biomass in the soil, *Soil Biol. Biochem.*, **11**, 147-148.

Swisher, R. and Carroll, G.C. (1980) Fluorescien diacetate hydrolysis as an estimation of microbial biomass on coniferous needle surfaces, *Microbial Ecology*, **6**, 217-226.

Thornton, C. (1996) Development of monoclonal antibody-based immunoassays for the quantification of *Rhizoctonia solani* and *Trichoderma harzianum* in soil, in this volume.

Whipps, J.M. (1987) Method for estimation of chitin content of mycelium of ectomycorrhizal fungi grown in solid substrates, *Trans. Brit. Mycol. Soc.*, **89**, 199-203.

West, A.W., Grant, W.D. and Sparling, G.P. (1987) Use of ergosterol, diaminopimelic acid and glucosamine contents of soils to monitor changes in microbial populations on solid substrates, *Trans. Brit. Mycol. Soc.*, **89**, 199-203.

SOME ASPECTS OF FUNGAL ECOLOGY

BENGT SÖDERSTRÖM
Lund University
Department of Microbial Ecology
Ecology Building, S-223 62 Lund
Sweden

1. Introduction

Fungi are dominant microorganisms in all terrestrial ecosystems. They are major saprophytes, they are very important pathogens, and they are key symbionts. The ecology of the fungi is thus of essential importance for the functioning of the whole ecosystem. In this contribution some very general aspects of fungal ecology will be shortly addressed with special emphasis on terrestrial, non-pathogenic fungi. In particular, fungal succession will be discussed, that is how different fungal species or groups of species or physiological groups follow each other in different substrata. When considering introduction of new fungi into an environment, be these genetically modified or not, knowledge on the successional patterns of the fungal community existing in the ecosystem in question is important for increasing the chances of a successful introduction. Some aspects on disturbance effects on the fungal community and function of the community of disturbances will also be considered. One example of a disturbance is the introduction of a new species, clearly illustrating the relevance of disturbance studies. For better understanding both successional and disturbance aspects of fungal ecology, some different life strategies of fungi will be referred to. More comprehensive presentations of these aspects of fungal ecology are given by e.g. Cooke and Rayner (1984), Frankland (1981, 1992), Zak (1992).

2. Fungal Succession

A fungal community is dynamic; population and community composition as well as function is changing qualitatively and quantitatively with time (Frankland, 1981). Succession is one important part of these dynamics. Succession can be defined as 'a directional change in the composition, relative abundance, and spatial pattern of species comprising a community' (Frankland, 1992). This definition is true for both plants and fungi. However, the basically indeterminate growth and spread by fungal mycelia makes the fungal succession different compared to plant successions, and Rayner and Todd (1979) narrowed the general definition to better apply specifically to fungi: 'the sequential

11

D. F. Jensen et al. (eds.), Monitoring Antagonistic Fungi Deliberately Released into the Environment, 11–16.
© 1996 *Kluwer Academic Publishers. Printed in the Netherlands.*

occupation of the same site by thalli (normally mycelia) either of different fungi or of different associations of fungi'.

One fundamental difference between plant and fungal succession is that plants tend to build up what is known as climax communities in a stable environment. The development to such communities tend to show a typical and predictable succession. Such a climax community is, among other things, characterized by a very tight and conservative nutrient circulation with minor nutrient leakages, often in combination with a high species diversity. This is a situation that has never really been demonstrated for fungi. Rather, an increase in diversity of fungal species in later phases of a succession tends to result in an increased nutrient (plant nutrient) release from the system. This likely mirrors the difference in fundamental resource demand within each group or organism - the classical plant nutrients are not what a fungal community has to optimise in the latter stages of a succession. Rather, it is the energy resource that is optimized. This lack of similarity between plant and fungal ecosystem development might of course also mirror our relative ignorance of the fungal system and the way to compare this to other organism systems.

About 25 years ago Park (1968) suggested a separation of fungal successions into two different types, a division that still is generally accepted. He described 'substratum type succession' and 'seral type succession'. Substratum succession, which with more modern terminology should be called resource succession, is the succession of species that occurs on any decomposable material and such successions has been described for some pure substances like cotton or cellulose film, as in the classical work on succession by Tribe (1957). Resource succession has been studied in a few natural resources, particularly during 1960 and 1970ties, and as good examples, Frankland's (1966, 1969, 1976) studies on the fungal succession on bracken *(Pteridium aquilinum)* and Kendrick and Burges (1962) on pine needles, may serve. The investigations on resource successions have often been able to demonstrate typical successional patterns, with features that can often be related to the physiological capacity of the different species. The successions are obviously accompanied by a degradation and subsequent chemical change of the resource, and, in temperate regions, after 5-6 years only very slowly mineralized material is left. Much work has also been done on woody resources (trunks and twigs) (Rayner and Boddy, 1988a).

Seral succession is the fungal succession occurring in a developing ecosystem, like a forest in a phase of establishment. The change in substrata presented to the decomposing community in the soil in such systems will result in a seral succession of fungi. Thus, the fungal seral succession follows the development of climax community of plants, and will of course be an integration of a number of resource successions. Indeed, however, few good studies on fungal seral succession have been published and this kind of succession is consequently less well described, probably mainly because such studies have to be truly long-term studies if one wants to follow a development of a climax ecosystem. The colonization of new sand dunes has been studied (Brown 1958) and the mount St Helens disaster and stripmining events has also been used for some seral successional descriptions

(Allen and Allen, 1980; Allen 1987). One recent and good example of a seral succession of fungi is the one described for ectomycorrhizal fungi (Last *et al.* 1984). Here the formation of mycorrhizal fruiting bodies was studied in a birch plantation on old farmland. No birch had been grown there before, and for the mycorrhizal fungi, the ecosystem was thus virgin. This particular study has in fact been interpreted as a general succession, but is better regarded as a good example of a seral succession.

A succession, be that resource or seral succession, is of course strongly influenced by which species are at place at different stages of the possible succession, and consequently, the dispersion potential of species is of importance. In addition, the physiological characters of the involved fungi are of great importance, in particular for the establishment and possible removal of other species upon arrival. For example, it is evident that a fungus with a lignin decomposing potential may be favoured in the later stages of a resource succession when lignin becomes a dominating chemical component of a litter. There are numerous examples in the literature to demonstrate this and similar situations.

A practical problem with most studies of succession is sampling and cultivation. As macro-organisms, we tend to sample macro-habitats rather that micro-habitats. Because of these course sampling procedures, we are often not really able to describe the natural patchiness which is immediately found when proper sampling techniques are applied, and consequently, the true successions can often not be distinguished. At the best, we may find an integrated successional pattern. Further, many studies confine themselves to fungi possible to grow in pure culture, often on far too rich media. Then, of course, a totally unrepresentative picture of a true succession will be obtained. Some of the recently developed molecular techniques will probably be very useful for a better description of the true successional patterns in natural resources.

Finally, it must be emphasized that we do still know very little about the true fungal successions in nature. Most studies have concentrated on succession of fruiting structures, or on fungi we are able to easily isolate. With the studies in wooden resources as a striking exception (Rayner and Boddy, 1988a, b), very little work has been done directly on the vegetative mycelia and the mycelial succession in nature.

3. Fungal Strategies

It is of course not easy to describe neither all mechanisms that influence the succession nor what governs these mechanisms. However, during the evolution, selection has resulted in a number of different life strategies and the different life strategies adopted by different fungal groups or species could be helpful to consider. One simple and very widely accepted classification of different strategies is in K and r selected organisms as two extremes. r-selected organisms are classified as those having a short individual life expectancy and which spend much of their effort in a reproductive effort. K-selected organisms are supposed to have a long individual life expectancy, and a relatively small effort put into reproduction. Most organisms fall between these two extremes. The model of plant strategies proposed by Grime (1979) was adopted to fungal strategies by Cooke and Rayner (1984) and this model can also be helpful when trying to explain an observed succession. They discuss three main strategies: R (ruderals), C (competitive), S (stress) strategies. R-selection, ruderal strategy, organisms have a short life span in combination with high reproductive potential giving success in disturbed but nutrient rich conditions. They are fast growers, have a high reproductive potential, quick spore gemination and growth. Some species within the Mucorales are typical for this group as well as the fungi establishing themselves after fires. We may recognize these as typical r-selected organisms. The C, combative (or competitive) strategy, organisms maximize occupation and exploitation of resources in a relatively nonstressed and undisturbed condition. They normally show a long term growth but they also have the potential of combat against other fungi. These organisms thus are typically K-selected, but their combative (defensive or aggressive) potential is also important. The S-selection, stress tolerant, strategy organisms have developed adaptations allowing survival under continued environmental stress. Stress may be nutrient based or other stresses. Typical S-selected groups are e.g. lignicoles, thermotolerant or xerotolerant fungi. Most fungi can be placed somewhere between these three basic strategies and comparing different fungi in respect to their life strategies might be helpful for better understanding their ecological potential.

4. Disturbances

Successions are of course also influenced by disturbances. But, more, whatever we do in nature, can be regarded as a disturbance. A disturbance may affect everything from the environmental heterogeneity to the temporal heterogeneity or modifying the relative abundance of species (Zak, 1992). An ecological disturbance has been defined as: "any relatively discrete event in time that disrupts ecosystem, community, or population structure and changes resources, substrate availability, or the physical environment" (White and Picket, 1985). It is evident that natural and unnatural disturbances may occur. A natural event like litter fall can be regarded as a disturbance, as well as a more dramatic unnatural event like stripmining.

Depending on the level of studies, a disturbance may be very differently interpreted. A disturbance may affect certain species and even change species diversity and species richness (Wicklow and Wittingham, 1974; Gochenauer, 1981). However, if the function of the new community can perform the same ecosystem functioning, one might claim the disturbance is not a disturbance from a functional point of view. Thus, it is important to define the level on which the disturbance is studied. We might also discuss large and small scale disturbances. Most work has been done on large scale disturbances like fire, a natural large scale disturbance, to anthropogenic large scale disturbances like agriculture or strip mining. Very little work has been done on small scale disturbances. Small scale disturbances of course are the most frequent ones in a microenviroment like the fungal environment. Small scale disturbances still may have very important ecological effects, quite different from large scale disturbances. For example, a large scale disturbance generally decreases the heterogeneity of the environment, while a small scale disturbance very well may increase the spatial and temporal heterogeneity of a system (Zak, 1992), and consequently, new patches for colonization may open up and an increased overall diversity and species richness might be the result. In fact, without any small scale disturbances, one could have expected a more homogenous fungal flora to develop and fungi with certain characters, and the C-strategy fungi would have outcompeted most other fungi. The disturbances keeps the ecosystem young and developing, but it is also generally believed that disturbance, stress and competition have been the major driving forces in the development of different survival strategies. However, for enabling interpretation of effects of disturbances, much more information is needed on small scale disturbances.

5. Conclusions

In agriculture and forestry, the beneficial potential of introducing new microorganisms, genetically modified or not, is great. However, before considering such introductions, understanding the basic ecology of the microorganism in the environment where the introduction is planned, is essential. With this knowledge, a better controlled introduction can be achieved thereby increasing the potential of a successful introduction. Understanding effects on the fungal community of introduced disturbances is one such basic desirable knowledge, and, in particular, knowledge of an ongoing course of succession and what factors govern the structure of the succession, might be of great help in planning a safe and successful introduction.

6. References

Allen M.F. (1987) Re-establishment of mycorrhizas on Mount St. Helens: migration vectors. *Transactions of the British Mycological Society* **88**, 413-417.

Allen E.B. and Allen M.F. (1980) Natural re-establishment of vesicular-arbuscular mycorrhizae following stripmine reclamation in Wyoming. *Journal of Applied Ecology* **17**, 139-147.

Brown J.C. (1958) Soil fungi in some British sand dunes in relation to soil type and succession. *Journal of Ecology* **46**, 641-664.

Cooke R.C. and Rayner A.D.M. (1984) *Ecology of Saprotrophic Fungi*. Longman, London.

Frankland J.C. (1966) Succession of fungi on decaying petiols of Pteridium aquilinum. *Journal of Ecology* **54**, 41-63

Frankland J.C. (1969) Fungal decomposition of bracken petiols. *Journal of Ecology* **57**, 25-36

Frankland J.C. (1981) Mechanisms in fungal succession. In *The Fungal Community: Its Organization and Role in Ecosystem* (eds. D.T. Wicklow and G.C. Carroll). Marcel Dekker: New York and Basel. pp 403-426.

Frankland J.C. (1992) Mechanisms in fungal succession. In *The Fungal Community. Its organization and Role in the Ecosystem*. 2nd Ed. (eds G.C. Carroll and D.T. Wicklow). M. Dekker Inc, New York. pp 383-401.

Gochenaur S.E. (1981) Response of fungal communities to disturbances. In *The Fungal Community: Its organization and Role in the Ecosystem* (eds D.T. Wicklow and G.C. Carroll). M. Dekker Inc, NY. pp 459-479.

Grime J.P. (1979) *Plant Strategies and Vegetation Processes*. John Wiley. Chichester and New York.

Kendrick W.B. and Burges A. (1962) Biological aspects of the decay of *Pinus silvestris* leaf litter. *Nova Hedwigia* **4**, 313-342.

16

Last F.T., Mason P.A., Ingleby K. and Fleming L.V. (1984) Succession of fruitbodies of sheathing mycorrhizal fungi associated with *Betula pendula*. *Forest Ecology and Management* **9**, 229-234.

Park D. (1968) The ecology or terrestrial fungi. In *The Fungi*, Vol 3 (eds G.C. Ainsworth and A.S. Sussman) Academic Press, New York. pp 5-39.

Rayner A.D.M. and Boddy L. (1988a) *Fungal Decomposition of Wood: Its Biology and Ecology*. John Wiley, Chichester, New York, Brisbane, Toronto, Singapore.

Rayner A.D.M. and Boddy L. (1988b) Fungal communities in the decay of wood. *Advances in Microbial Ecology* **10**, 115-136.

Rayner A.D.M.and Todd N.K. (1979) Population and community structure and dynamics of fungi in decaying wood. *Advances in Botanical Research* **7**, 333-420.

Tribe H.T. (1957) Ecology of micro-organisms in soil as observed during their development upon buried cellulose film. In *Microbial Ecology* (eds R.E.O. Williams and C.C. Spicer). Cambridge University Press, New York. pp 287-289.

White P.S. and Picket S.T.A. (1985) Natural disturbances and patch dynamics: an introduction. In *The Ecology of Natural Disturbances and Patch Dynamics* (eds S.T.A. Picket and P.S. White). Academic Press, New York. pp 139-194.

Wicklow D.T. and Wittingham W.F. (1974) Soil microfungal changes among the profiles of disturbed conifer-hardwood forests. *Ecology* **55**, 3-16.

Zak J.C. (1992) Response of soil fungal communities to disturbances. In *The Fungal Community. Its Organization and Role in the Ecosystem*. 2nd Ed. (eds G.C. Carroll and D.T. Wicklow). M. Dekker Inc, New York. pp 403-425.

QUANTIFICATION OF FUNGAL GROWTH IN THE ENVIRONMENT

SCHNÜRER, J. [1] AND BÖRJESSON, T.[2]

[1] Department of Microbiology, Swedish University of Agricultural Sciences, Box 7025, S-750 07 Uppsala, Sweden. [2] Swedish Farmers Supply and Marketing Association, Box 30192, S-104 25 Stockholm, Sweden

Abstract

New and highly specific assays for monitoring antagonistic fungi in the environment always have to be calibrated aginst " base" methods for quantification of fungi. These methods should either determine the total amount of fungal biomass in a sample or quantitatively determine the presence of certain species or genera. Examples of such "base" methods are the quantification of colony forming units (CFU), hyphal lenghts and ergosterol. Development of selective substrates can increase the precision of CFU determinations, as has been shown in food mycology. However, results are to a large extent only a measure of the degree of fungal sporulation. Direct microscopy for hyphal lenght (and spore number) detemations have been much used in soil microbiology. In combination with good staining techniques, e.g. immuno-fluorescence, reliable determinations of fungal biomass can be obtained even in difficult samples. The main drawback is that the procedure is very tiresome for the eyes of the operator. Subsequently only limited number of samples can be handled. Quantification of chemical components of fungi, such as chitin or ergosterol, can be used with large numbers of samples. Ergosterol is now a widely used marker of the total amount of fungal biomass, especially in food mycology and plant pathology. This lipid is found in membranes of fungi in Deuteromycotina, Ascomycotina, Zygomycotina and Basidiomycotina. In pure culture experiments ergosterol has been found to be highly correlated to mycelial dry weight, mycelial protein, hyphal lengths, chitin, CO_2-evolution, volatile fungal metabolites, mycotoxins, immunological markers and sometimes with levels of fungal colony forming units. Results from determinations of ergosterol can for a single sample be obtained within an hour. However, there is a need for even faster methods. Near infrared reflectance spectroscopy (NIR) offers fast quantification of chemical constituents of opaque biological materials, such as protein in cereals. We have found that (NIR) spectra correlates with ergosterol and can be used for a rapid prediction of fungal biomass. Sensor technology offers other possibilities. We have found that an " electronic nose", where signals from three different type of gas-sensitive sensors are processed with an artifical neural network, can be trained to quantitatively predict ergosterol contents of plant material. Volatile fungal metabolites thus react with the sensors and form signal patterns that indicate fungal biomass both qualitatively and quantitatively. Volatile patterns from different *Penicillium* species can also be differentiated by the "electronic nose".

Introduction

Monitoring of fungi in the environment requires methods able to detect and identify fungi

17

D. F. Jensen et al. (eds.), Monitoring Antagonistic Fungi Deliberately Released into the Environment, 17–24.
© 1996 Kluwer Academic Publishers. Printed in the Netherlands.

18

at different levels of specificity. That is either as fungi as such, as an ecological group, e.g. storage moulds or at the genus, species or strain level. Methods are also needed that can provide quantitative data on fungal mass and activity. Figure 1 summarises the status of commonly used methods with regard to specificity and ability to provide quantitative data. Methods based on DNA/RNA are the most specific, but have not yet been fully developed to provide quantative data. Hyphal length (volume) determinations through direct microscopy provides the most quantitative data (Schnürer, 1993) but does not provide a high degree of specificity, unless coupled to immunostaining techniques (Newell, 1992). New and highly specific assays for monitoring antagonistic fungi always have to be calibrated aginst "basal" methods for quantification of fungi. Examples of such "basal" methods are the quantification of colony forming units (CFU), hyphal lenghts and ergosterol.

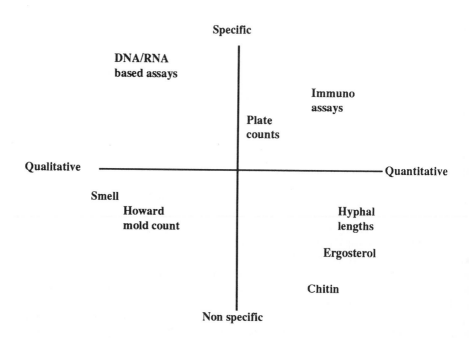

Figure 1. Representation of the relation between specificity and ability to provide quantitative data for methods used in varous fields of mycology.

COLONY FORMING UNITS

Development of selective substrates for various ecological groups can increase the precision of CFU determinations. In food mycology substantial progress has been made in this field (Samson *et al.*, 1995). Fungi with increasing degree of xero-tolerance can be detected on DG18 (Dichloran Glycerol 18 %) with a water activity of 0.95 or on MY 50G (Malt Yeastextract 50% Glucose) with water activity 0.89. Likewise "acido tolerant" fungi, e.g.

Penicillium roqueforti and preservative resistant yeasts, are among the few fungi able to grow on malt extract agar with 0.5 - 1.0 % acetic acid. Colour changes of the colony reverse, based on reactions between substrate components and secondary metabolites, indicate the presence of certain mycotoxigenic fung, e.g. *Aspergillus flavus* on AFPA and *Penicillium verrucosum* on DRYES (Dichloran Rose bengal Yeast Extract Sucrose). However, it has to be kept in mind that the results of fungal CFU determinations are to a large extent only a measure of the degree of fungal sporulation (Schnürer, 1993).

HYPHAL LENGTH

Direct microscopy for hyphal lenght (and spore number) deteminations have been much used in soil microbiology as well as in studies of litter decomposition. This provides a determination of fungal volume which can be converted to fungal (bio)mass. In combination with good staining techniques, e.g. immuno-fluorescence, reliable determinations of fungal biomass can be obtained even for difficult samples. Newell (1992) has written an excellent review on advantages and problems of using this and other methods of estimating fungal biomass in decomposing litter. A main drawback of direct microscopy is that the procedure is very tiresome for the eyes of the operator. Subsequently only limited numbers of samples can be handled.

ERGOSTEROL

Fungal specific marker substances, such as ergosterol and chitin, offers a third possibility for quantification of fungal mass. Of these the determination of ergosterol, originally proposed by Seitz *et al.* (1977) for determining the degree of fungal infestation of grains, have become increasingly used. Ergosterol is a dominant membrane sterol in all eumycotan fungi, except chytrids, rusts and some yeasts, but is absent or very rare in oomycetes and hyphochytridiomycetes (Newell, 1992). It is not found in bacteria, plants and animals, but has been found in certain algae and protozooa (Newell, 1992). The position of double bonds at position 5, 7 leads to a highly specific ultraviolet absorption spectrum with a maximum at 282 nm. This makes possible a quantification after separation by thin-layer chromatography (Naewbaniij *et al.* ,1984) or high pressure liquid chromatography (Seitz *et al.* ,1977; Newell *et al* 1988, Schwadorf and Müller, 1989). Fungal ergosterol contents normally range between 0.2 to 0.6 % of dry weight, but the content can vary more widely depending on mycelial age and growth conditions (Newell, 1992; Schnürer, 1993). Ergosterol contents have been found to correlate with chitin (Seitz *et al.* 1979), deoxynivalenol (Young *et al.* , 1984), mycelial protein (Zill *et al.,* 1988), volatile fungal metabolites (Börjesson *et al* ., 1990), colony forming units (Schnürer and Jonsson, 1992), mycelial surface antigens (Schwabe *et al.,* 1992) and with hyphal lengths (Schnürer, 1993).

In France quantative determinations of ergosterol are now routinely done to evaluate the mycological quality of feedstuffs and the analytical procedure has been standardized as French standard AFNOR V18-112 (Maupetit *et al.,* 1993). Müller *et al.* (1993) has during a six-year period determined ergosterol and myxotoxin levels in grain from South Western Germany (Baden-Würtenberg). These data have been used to construct a mathematical model of the relationship between ergosterol contents and the probability of exceeding toxicologically unacceptable levels of certain mycotoxins. Based on this model, data on ergosterol levels of grains forms the basis for advice given to farmers on how to use mould infested grain for different purposes (H-M. Müller, per. comm.)

Ergosterol determinations using the extraction-HPLC method requires approximately 90 minutes for a single sample. As an alternative we are presently investigating the use of near-infrared spectroscopy (NIR) for a rapid quantification of fungal mass in opaque biological material. Preliminary results suggests that NIR spectra can be used to predict ergosterol contents in milled grain samples and that results can be obtained within five minutes (Schnürer, J and Nilsson M, unpublished results).

The ergosterol assay has now become an established method for the quantification of fungal mass in food and feeds, especially in grains, but it has also been widely used for quantification of fungal mass in litter and even in soil (Newell 1992). However, it has to be remembered that the assay is non-specific and does not provide any information on species composition. Further, expensive equipment such as HPLC is needed for the standard procedure. The possibility of using NIR technology only aggrevates this problem. The solution might be the development of immunoassays specific to ergosterol. In spite of these potential drawbacks determinations of ergosterol can be used to detect fungal presence in many environments, as well as providing quantitative data on fungal mass.

ELECTRONIC NOSE

The most commonly used instrument to detect fungal growth is the human nose. It is used to detect mould and other objectionable odours of grains in all international, and in most national, trade. This method, although fast and sensitive, have a number of drawbacks. Most important among these is that the method is subjective. The inhalation of fungal spores from mouldy grain may also be hazardous (Rylander, 1986). There is thus a need to develop alternative methods that are efficient, not too expensive and preferably as fast as that of human inspection. The use of an electronic nose could meet these criteria. An electronic nose, employing sensor detection of volatiles trapped in a cold-trap, have earlier been constructed to classify grains (Stetter et al., 1993). We have earlier developed an electronic nose that does not require a cold-trap and that is able to classify grain according to odour characteristics (Jonsson et al., In press). We have also investigated whether the electronic nose can provide quantitative esitimates on ergosterol and whether it can be used to differentiate between fungal species.

Experimental procedure
The set up of the electronic nose is an automatic sampling apparatus, which can be loaded with 30 grain samples. Samples are heated to 65° C in a heating unit and headspace samples are pumped to an array of 15 sensors mounted in chambers. The sensors used are MOSFET (Metal Oxide Field Effect Transistor)-sensors constructed at the Laboratory of Applied Physics, Univerisity of Linköping, Sweden, Tin-oxide sensors (Taguchi-sensors, Figaro Engineering Inc., Japan) and an IR-detector monitoring carbondioxide (Edinburgh sensors, Edinburgh, Scotland). MOSFET-sensors are primarily sensitive to hydrogen containing compounds and their characteristics can be altered by changing the thickness and type of metal layer and operating temperature. Tin-oxide sensors are primarily sensitive to

combustible compounds. The responses from the sensors are treated with an Artificial Neural Network (ANN, Neural Network Explorer, Neural Ware, Pittsburgh, PA, USA). The ANN works in a similar manner as the human brain. It is able to connect an input pattern, such as responses from sensors, with a certain output, such as mouldy or sound grains (Figure 2). A large number of patterns from known samples are used to train the neural network during the training phase, which is followed by a test of the ability to classify unknown samples. Two thirds of the patterns are used as training set and one third as test set.

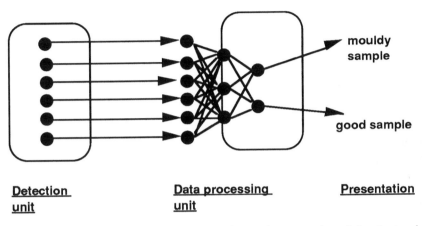

| **Detection unit** | **Data processing unit** | **Presentation** |

Figure 2.Schematic representation of data processing and presentation of the electronic nose.

Ergosterol quantification. To explore the possibility of predicting ergosterol content in grains, 30 barley samples with known ergosterol content were run in the electronic nose. Two third of these were replicated once, giving a total of 50 runs. The samples were divided into three groups having low (< 7 µg/g dw), medium (7-10 µg/g dw) and high (> 10 µg/g dw) ergosterol content.

Fungal species differentiation. The ability of the electronic nose to distinguish between different *Penicillium* species was investitigated in an *in vitro* experiment. Eighteen glass vessels (250 ml) containing MEA (2 %) were inoculated with *Penicillium glabrum* (SLU J3, 6 vessels), *Penicillium roqueforti* (SLU J5, 6 vessels) or left uninoculated (6 vessels). Spore suspensions containing 5×10^6 spores/ml were prepared and 0.2 ml were spread over the agar surface. Four days after inoculation, headspace samples were analysed with the electronic nose. Data was reduced to three patterns for each treatment, by grouping vessels together two by two. The means were then used for data processing.

Results
When memberships to three classes of ergosterol levels was predicted no more than 65 % of the barley samples were correctly classified (Table 1). However, no overlap between the classes having the highest and the lowest values was observed which suggests screening possibilities.

Table 1. Prediction of barley ergosterol content using the electronic nose. Three classes were constructed: Low (< 7 μg ergosterol / g dw) , medium (7-10 μg ergosterol / g dw) and high (< 10 μg ergosterol / g dw). Results are given as degre of connection with the three classes where 1 indicate full and 0 no connection.

		Prediction		
Actual class	Low	Medium	High	Correct?
Low	0.01	0.86	0.09	no
Low	0.81	-0.1	0.06	yes
Low	0.01	1.02	-0.05	no
Low	1.06	-0.12	-0.02	yes
Low	0.40	0.29	0.21	yes
Medium	0.11	0.87	0.01	yes
Medium	0	1.01	-0.04	yes
Medium	0	1.04	-0.05	yes
Medium	0.66	-0.07	0.17	no
Medium	0.88	-0.09	0.02	no
Medium	0	0.01	0.98	no
Medium	0.01	0.55	0.37	yes
High	0.01	0.26	0.69	yes
High	0.09	-0.03	0.83	yes
High	0	0	0.98	yes
High	-0.01	0.32	0.67	yes
High	0	1.04	-0.06	no

After four days of fungal growth, the electronic nose could correctly classify patterns from vessels inoculated with *P.glabrum, P. roqueforti* and uninoculated vessels (Table 2).

Table 2. The ability of the electronic nose to distinguish between volatiles from *Penicillium glabrum*, *Penicillium roqueforti* and uninoculated malt extract agar. Results are given as degre of connection with the three classes where 1 indicate full and 0 no connection

	Prediction		
Actual species	Uninoculated	*P. glabrum*	*P. roqueforti*
Uninoculated	0.96	0	0.02
P. glabrum	0	0.94	0.06
P. roqueforti	0.03	0	1

Conclusions

New technology, based on spectroscopic techniques and electronic sensors, can be used to detect and quantify mould growth. In the future the performance of the electronic nose will be of interest to compare NIR for fungal quantification. The ability of the electronic nose to correctly classify agar cultures of *P. glabrum* and *P. roqueforti* further suggests the possibility of using sensor technology to monitor specific antagonistic fungi in various environments.

References

Börjesson, T., Stöllman, U. and Schnürer, J. 1990. Volatile metabolites and otherindicators of *Penicillium aurantiogriseum* growth on different substrates. Appl. Environ. Microbiol. 56:3705-3710

Jonsson, A., Winquist, F., Schnürer, J., Sundgren, H. & Lundström, I. *In Press*. Electronic nose for microbial classification of grains. Food Microbiol. 00:000-000.

Maupetit, P, Gatel, F., Cahagnier, B., Botorel, G., Charlier, M. Collet, B., Dauvillier, P., Laffiteau, J. and Roux, G. 1993. Quantitative estimation of fungal infestation of feedsstuffs by determining ergosterol content. 44 th Annual Meeting of the EAAP, Aarhus Denmark, 16-19 Augusti, 1993.

Müller, H-M, Reimann, J., Schwadorf, K. and Thöni, H. 1993. Zur bewertung des Ergosteringehaltes von Futtermitteln. VDLUFA Schriftenreike, Heft 37, Kongressenbaud Hamburg.

Naewbanij, M., Seib, P.A., Burroughs, R. Seitz, L.M. and Chung, D.S. 1984. Determination of ergosterol using thin-layer chromatography and ultra-violet spectroscopy. Cereal. Chem. 61:385-388.

Newell, S.Y., Arsuffi, L.T. and Fallon, R.D. 1988. Fundamental procedures for determining ergosterol content in decaying plant material by liquid chromatography. Appl. Environ. Microbiol. 54:1876-1879.

Newell, S.Y. 1992. Estimating fungal biomass and productivity in decomposing litter. *In* Carroll, G.C. and Wicklow, D.T. (eds.) The Fungal Community - Its organisation and role in the Ecosystem, 2nd ed. Marcel Dekker, Inc. New York, Basel, HongKong. pp 521 - 561.

Rylander, R.1986. Lung diseases caused by organic dusts in the farm environment . American J. Ind. Med. 10:221-227.

Samson, R.A., Hoekstra, E.S., Frisvad, J.C. and Filtenborg, O. 1995. Introduction to Food-borne Fungi. Centraal-bureau voor Schimmelcultures. Baarn and Delft. 322 pages.

Schnürer, J. 1993 Comparision of methods for estimating the biomass of three food borne fungi with different growth patterns. Appl. Environ. Microbiol. 59:552-555.

Schwabe, M., Kamphuis, H., Trümner, U., Offenbächer, G. and Krämer, J. 1992. Comparison of the latex agglutination test and the ergosterol assay for the detection of moulds in foods and feedstuffs. Food. Agric. Immunol. 4:19-25.

Schwadorf, K. and Müller, H-M. 1989. Determination of ergosterol in cereals, mixed feed components and mixed feeds by liquid chromatography. J. Ass.Off. Anal. Chem. 72:457-462.

Seitz, L.M., Mohr, H.E., Burroughs, R. and Sauer, D.B. 1977. Ergosterol as an indicator of fungal invasion in grains. Cereal Chemistry 54:1201-1217.

Seitz, L.M., Sauer, D.B., Burroughs, R., Mohr, H.E. and Hubbard, J.D. 1979. Ergosterol as a measure of fungal growth. Phytopathology 69:1202-1203

Stetter, J.R., Findlay, M.W., Schroeder, K.M., Yue, C. & Penrose, W.R. 1993. Quality classification of grain using a sensor array and pattern recognition. Anal. Chem. Acta. 284:1-11.

Young, J.C. Fulcher, R.G., Hayhoe, J.H., Scott, P.M. and Dexter, J.E. 1984. Effect of milling and baking on deoxynivalenol(vomitoxin) content of estern Canadian wheats. J. Agric. Food Chem. 32:659-664.

Zill, G., Engelhardt, G. and Wallnöffer, P.R. 1988. Determination of ergosterol as a measure of fungal growth using Si 60 HPLC. Z. Lebensm. Unters. Forsch. 187:246-249.

POPULATION GROWTHS AND SURVIVAL OF *TRICHODERMA HARZIANUM* AND *TRICHODERMA VIRENS* IN SPHAGNUM PEAT.

NINA HEIBERG[1], HELGE GREEN AND DAN FUNCK JENSEN

Department of Plant Biology, Plant Pathology Section, The Royal Veterinary and Agricultural University, Thorvaldsenvej 40, DK-1871 Frederiksberg C, Copenhagen, Denmark.

[1] *New address: The Norwegian Crop Research Institute, Ullensvang Research Centre, Division Njøs, N-5840 Hermansverk, Norway.*

1. Abstract

Survival and population development of antagonistic strains of *Trichoderma harzianum* (T3) and *Trichoderma virens* (G2), were studied separately in pot experiments using a sphagnum peat mixture as growth substrate. The population density was measured by dilution plating on selective media. Activation and germination of *T. harzianum* (T3) conidia were studied in the growth substrate using FDA-staining.

In steamed sphagnum peat, population density recorded as colony forming units decreased from day 0 to day 2, indicating that a large amount of the conidia had germinated. Results from FDA-staining confirmed that the number of active and germinated conidia of T3 were highest after two days. The number of cfu of both antagonists increased up to day 32, from where it started to decrease, but most of the conidia were still viable after 64 days. In unsteamed sphagnum peat no significant changes in population density was observed, during a period of 64 days. Although scattered swelling, germination and formation of conidia seemed to take place at any time during the 64 days, this had no detectable influence on the cfu. Rather, it indicates a constant turnover of the total biomass, which is limited by the amount of available nutrients.

D. F. Jensen et al. (eds.), Monitoring Antagonistic Fungi Deliberately Released into the Environment, 25–31.
© 1996 *Kluwer Academic Publishers. Printed in the Netherlands.*

2. Introduction

The fungal antagonists *Trichoderma harzianum* and *Trichoderma virens (Gliocladium virens)* have been shown to be effective biocontrol agents against a range of soil-borne pathogens (Papavizas, 1985). For effective biocontrol of soilborne pathogens it is of major importance that the antagonists are able to be established and proliferate after introduction into soil.

The present experiments were carried out in order to study the survival and proliferation of the antagonists after their incorporation in sphagnum peat.

3. Material and Methods.

Strains of *Trichoderma harzianum* (T3) and *T. virens* (G3), both isolated from *Pythium*-suppressive peat (Wolffhechel, 1989) were maintained on potato-dextrose-agar (PDA). Peat bran inocula of the fungi were prepared according to Sivan *et al.* (1984).

Fine 0-20 mm, light coloured sphagnum peat (Pinstrup, faerdigblanding 1, Denmark) mixed with vermiculite, grade 11 (Skamol, Denmark), 3:2 (w:w), pH (H₂O) 5.5 were used as growth substrate. Both steamed- and unsteamed growth substrate were included in the experiments and steaming was carried out for one hour on three successive days.

Peat bran inoculum was thoroughly mixed with the growth substrate. As the amount of inoculum of T3 added had no significant influence on the final population density in steamed sphagnum peat one or two weeks after incorporation (results not shown), 0.1 % vol. was chosen for the following experiments. The substrate was filled into 330 ml pots giving 27.5 g dry weight of growth mixture per pot. Each pot were watered with a weak balanced nutrient solution to a final water content of 64 % vol and a conductivity of 3.5 mMho. The pots were placed in a growth chamber at 18°C with a 16 hours photoperiod.

In the experiment samples were taken on the day of mixing, and then on day 2 and 4, with three replicates per treatment. In the second experiment, samples were taken on day 2, 4, 8, 16, 32 and 64, with two replicates per treatment. In this experiment the pots were kept in plastic bags in the growth chamber and watered to the initial weight once a week. Population density of the antagonists were determined by dilution-plating on *Trichoderma*-selective medium (Elad *et al.*, 1981), and expressed as colony forming units (cfu) per gram of oven-dried (86°C) growth mixture. The plates were incubated for 4 days at room temperature prior to recording.

FDA-staining of conidia and hyphae was carried out to study the actual growth phases of *T. harzianum* (T3). The first sample were taken on the day of mixing and then on day 2, 4, 7, 14, 21 and 64. Approximately 1.4 g (dry weight) of the sphagnum was sampled, homogenized in 100 ml 50mM Phosphate buffer, pH 7.0, diluted 1:100, and 10 μg/ml fluorescein diacetate (FDA, Koch-Light Laboratories) in acetone was added. After 5 minutes, 5 ml of the suspension was filtered through a black 8 μm, polycarbonate membrane filter (Nuclepore, Costar Scientific Corporation) and placed in a drop of 2.5 % (v/w) 1,2-diazyabicyclo- [2,2,2]-octane (Sigma D-2522) in phosphate buffer to reduce fading. The slide was viewed under a Nicon Optiphot epi-fluorescense microscope equipped with an excitation filter at 430-484 nm and a barrier at 520 nm. Between 100-200 conidia were examined.

4. Results

In steamed sphagnum peat, the population density of *T. harzianum* (T3), recorded as colony forming units decreased approximately to one third from day 0 to day 2 (Figure 1.). Thereafter an increase in population density was observed. The population density of *T. virens* (G2) was smaller, but the same decrease in population density from day 0 to day 2 was observed (Figure 2.). From day 2 the population density of both antagonists rapidly increased and reached a maximum at day 32, but most of the conidia were still viable after 64 days (Figure 3. and 4.).

In unsteamed sphagnum peat only a small and insignificant decrease in population density was observed from day 0 to day 2 (Figure 1. and 2.). No significant changes in population density were observed for any of the antagonists in unsteamed sphagnum peat from day 2 to day 64 (Figure 3. and 4.).

Results from FDA-staining showed that the numbers of active and germinated conidia of T3 in steamed sphagnum peat were highest after two days (Figure 5.). On the second day 20 % of the conidia were resting while 50 % had become active (swollen) and 30% germinated (Figure 2.). On day 4, when formation of conidia had started, the percentage of active conidia were 53% but, although the germ tubes were much longer, the number of germinated conidia had decreased to 10%. On day 7 less than 20% of the spores were swollen or had germinated, and on day 14, less than 1% of the conidia were active and only sparse germination occurred. Resting conidia measured 2.97 X 2,69 μm and reached a size of 5.81-7.74 μm in active state.

In unsteamed sphagnum much fewer active conidia were present. Results from FDA-staining showed that within the first two weeks, maximum was reached the second day with 18 % swollen and germinated conidia.

28

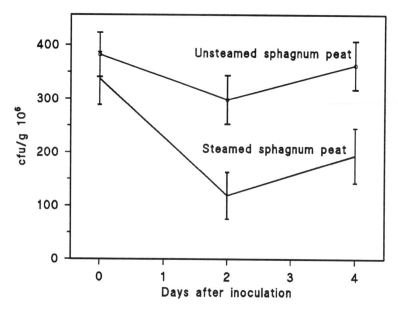

Figure 1. Population development of *T. harzianum* (T3) the first days after inoculation of the growth substrate measured by dilution plating and expressed as cfu per gram dried sphagnum peat.

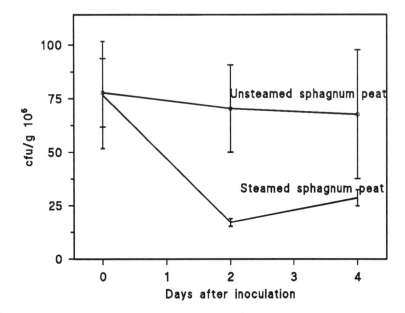

Figure 2. Population development of *T. virens* (G2) the first days after inoculation of the growth substrate measured by dilution plating and expressed as cfu per gram dried sphagnum peat.

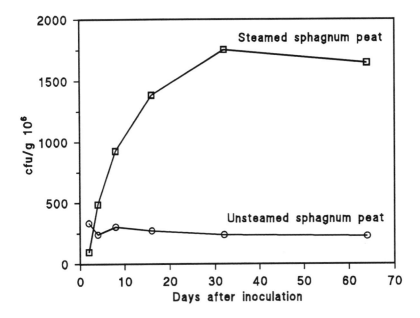

Figure 3. Population development of *T. harzianum* (T3) measured by dilution plating and expressed as cfu per gram dried sphagnum peat.

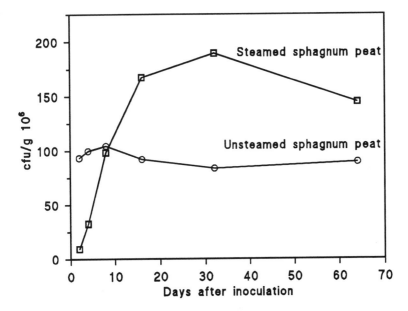

Figure 4. Population development of *T. virens* (G2) measured by dilution plating and expressed as cfu per gram dried sphagnum peat.

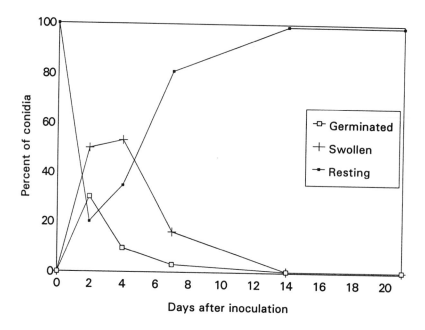

Figure 5. Distribution between resting, active (swollen) and germinated conidia of *T. harzianum* (T3) in the moist sphagnum peat. The conidia were made visible by FDA-staining. Observations on day 64 are excluded in the Figure, as no changes were observed from day 21 to day 64.

5. Discussion

Papavizas (1981) found that survival of *T. harzianum* conidia added to soil depended on the isolate used, but all isolates lost 50% or more of the original number of colony forming units (cfu) added after 35 days. After 75 days the population density dropped to less than 30% and after 130 days to less than 10% for all isolates.

Most of the conidia of T3 and G2 were still viable after 64 days. This indicate that conidia of T3 and G2 are able to survive for a long period in sphagnum peat. But most of the germination took place during a short period after incorporation, and a high level of germination only occurred in steamed sphagnum peat. In unsteamed sphagnum peat only a small percentage of the conidia germinated. Although scattered swelling occured, germination and formation of conidia seemed to take place at any time during the 64 days and this had no detectable influence on the cfu. Rather, it indicates a small but constant turnover of the total biomass.

The steaming process kills microorganisms in the sphagnum peat and as a result nutrients are released. As the amount of inoculum applied did not have significant influence on the final population density in steamed sphagnum peat, it is most likely that the amount of nutrients available was the limiting factor for proliferation and sporulation of the antagonist in steamed sphagnum peat. In unsteamed sphagnum peat, competition from other microorganisms might also have been a limiting factor.

The dilution plating technique has been widely used for quantitative isolation of antagonists from soil (Elad et al., 1981). But as the majority of colonies arise from conidia rather than hyphae (Warcup 1955), the method is mostly suitable for quantification of the total population of conidia in soil or growth substrates, rather than assessing the activity of the fungi. The drop in number of cfu recorded at the first few days of the present experiment, indicate that swollen (active) and germinated conidia are vulnerable to the dilution plating technique. As such, most of the cfu recorded are believed to originate from resting conidia. However, growth cycles result in formation of new conidia, and if sufficient nutrient are available, this can be seen as an increase in cfu.

6. References

Elad Y, Chet I and Henis Y. (1981) A selective medium for improving quantitative isolation of *Trichoderma* spp. from soil. *Phytoparasitica* **9**, 59-67.

Papavizas G.C. (1981) Survival of Trichoderma harzianum in soil and in pea and bean rhizospheres. *Phytopathology* **72**, 121-125.

Papavizas G.C. (1985). *Trichoderma* and *Gliocladium*: Biology, ecology, and potential for biocontrol. *Ann. Rev. Phytopathol.* **23**, 23-54.

Sivan A., Elad, Y. and Chet, I. (1984). Biological control effects of a new isolate of *Trichoderma harzianum* on *Pythium aphanidermatum*. *Phytopathology* **74**, 498-501.

Warcup, J.H. (1955). On the origin of colonies of fungi developing on soil dilution plates. *Trans. Br. Mycol. Soc.* **38**, 298-262.

Wolffhechel H. (1989) Fungal antagonists of *Pythium ultimum* isolated from a disease-suppressive Sphagnum peat. *Växtskyddnotiser* **53**, 7-11.

OCCURRENCE OF *GLIOCLADIUM ROSEUM* ON BARLEY ROOTS IN SAND AND FIELD SOIL.

INGE M.B. KNUDSEN, BIRGIT JENSEN, DAN FUNCK JENSEN AND JOHN HOCKENHULL.
Plant Pathology Section, Department of Plant Biology, The Royal Veterinary and Agricultural University, Thorvaldsensvej 40, DK-1871 Frederiksberg C., Copenhagen, Denmark.

1. Abstract

The rhizosphere competence on barley of the antagonist, *Gliocladium roseum* (isolate IK 726), was investigated in semi-sterile sand and field soil. Following application of the antagonist to the seeds it was shown by the dilution plating method, that *G. roseum* was present on the roots. In sand colonization occurred most on proximal compared to distal parts of the roots. In field soil, 4 month after sowing, *G. roseum* was recovered from roots of plants derived both from seed inoculated with *G. roseum* and non inoculated with the antagonist. However, the population of *G. roseum* was significantly higher on roots derived from inoculated seeds.

2. Introduction

Fungal colonization of the rhizosphere of developing roots or rhizosphere competence (as defined by Baker, 1991) is important for the success of a biocontrol agent since not only the seed may be protected from soilborne plant pathogens but also the roots as well.

Although antagonists applied to seeds have the opportunity to be primary colonizers of roots, there are several reports of seed inoculants failing to prevent root diseases (Kraft & Papavizas, 1983; Merriman *et al.*, 1974). This is perhaps because the tested organisms were poor surface colonizers (Windels, 1981) and/or because they were incapable of being transported by the root through the soil profile (Madsen & Alexander, 1982). Little information is available regarding movement of microorganisms from the seed to the rhizosphere, but even 'rhizosphere competent' microorganisms benefit from percolating water for their spread down roots (Chao *et al.*

33

D. F. Jensen et al. (eds.), Monitoring Antagonistic Fungi Deliberately Released into the Environment, 33–37.

1986; Krauss & Deacon, 1994). Thus biocontrol organisms with the ability to produce an abundance of free wettable propagules (including spores) in the rhizosphere may be more effective rhizosphere inhabitants. Chao *et al.* (1986) reported that while plant roots growing from soil into soil-free moist chambers were not colonized by *Trichoderma* spp., the fungi were able to colonize the upper half of the roots in sterile soil. These results indicate that transportation of *Trichoderma* along the roots is mainly a passive process.

The habitat (the source) of an antagonistic isolate may be important for its rhizosphere competence. Older literature indicates that bacteria and fungi colonizing seeds may be poor root colonizers (Peterson, 1959; Rouatt, 1959). The isolate of *Gliocladium roseum* used in this study, IK 726, was isolated from roots of barley seedlings infected with *Fusarium culmorum*. In field experiments (Knudsen *et al.*, 1995), this isolate has given biological control of seed-borne *F. culmorum* and *Bipolaris sorokiniana* at least as good as standard chemical treatments. In culture, *G. roseum* produces numerous conidia and may be comparable to *Trichoderma* regarding movement of conidia from the seed to the root surface (Chao *et al.*, 1986). As *G. roseum* is also known to be a root surface inhabitant (Pugh, 1980), good rhizosphere competence is to be expected.

The aim of our experiments was to investigate whether *G. roseum* applied to the seed was able to colonize root-tips of barley in a semi-sterile environment (sand) and whether seed treatment with *G. roseum* resulted in root colonization under field conditions.

3. Methods

3.1. SEMI-STERILE SAND EXPERIMENTS

Seeds of barley ('Alis') naturally infected with *B. sorokiniana* were treated with *G. roseum* (IK 726) in a 2% Pel gel[R] spore suspension (10^6 spores per ml). Controls consisted of seed treated with Pel gel[R] without *G. roseum*. The test was carried out at 15°C in small pot-strips containing 6 pots with moistened sand. The sand was unsterilized builder's washed sand, grade 0.4-0.8 mm, pH 6.42 - 6.5. Three seeds were sown in each pot and the pots were arranged in a randomized block design with 4 replicates.

After 19 days, plants were gently shaken free of sand and roots of 18 plants per replicate were sampled. Segments (0.5 cm) from the distal and the proximal ends of the roots were excised, weighed and washed separately in 20 x their measured weight of sterile distilled water at 1500 rpm (IKA-Vibrax VX 7) for 10 min. 0.1 ml aliquots of the washings were plated on *Trichoderma*-selective agar (Elad & Chet, 1983) in 3 replicates diluted to 2×10^4. The cfu of *G. roseum* per g root was calculated.

3.2. FIELD SOIL EXPERIMENTS

Seeds of barley ('Alis') artificially inoculated with *F. culmorum* (Knudsen *et al.*, 1995), were treated with a spore suspension (10^7 cfu per ml) of *G. roseum* (IK 726). The number of cfu per seed was approximately 10^4. Both treated and untreated seeds were sown in field plots and the plants were dug up and carefully separated from the bulk soil. Four month after sowing, samples consisting of entire roots, 5 g per replicate, were washed in 100 ml sterile distilled water according to the procedure outlined above. The suspensions were diluted to 2×10^{-4} and plated on *Trichoderma*-selective agar as described above. The cfu of *G. roseum* per g root was calculated.

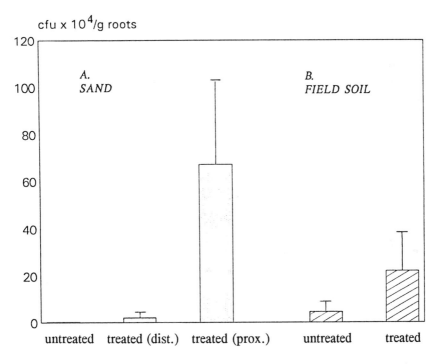

Figure 1. Number of cfu of *Gliocladium roseum* per g root: A: Isolation from distal or proximal segments of roots from plants derived from infected (*Bipolaris sorokiniana*) seed untreated or treated with *G. roseum* and grown in sand; B: Isolations from from complete roots from plants derived from inoculated (*Fusarium culmorum*) seed untreated and treated with *G. roseum* and grown in field soil.

4. Results

Distal segments of roots in sand yielded 2.5 x 10⁴ cfu of *G. roseum* per g root (Fig. 1) while, significantly higher (about 25 x) cfu counts per g root was recovered from the proximal root segments. No colonies of *G. roseum* were recovered from roots of the control plants.

Agar plates containing dilutions of washings from roots grown in field soil were heavily contaminated with *Pencillium* and *Cladosporium* which masked the colonies of *G. roseum* at the lower dilutions. Only at dilutions of 10⁻³ and 2x10⁻⁴ was it possible to differentiate colonies of *G. roseum* and these data was used to calculate the cfu per g root. It was found that roots from plants derived from seeds treated with *G. roseum* yielded 2.4 x 10⁵ cfu per g root compared to 4.4 x 10⁴ cfu per g root from plants derived from seeds not treated with *G. roseum* (Fig. 1).

5. Discussion

Stenton (1958) proposed that within the soil the newly emerging roots of higher plants provide a "virgin ecological niche to soil microorganisms". Thus, initially, such roots may be regarded as a substrate with little or no fungal colonization. As the root grows through the soil it releases exudates, which, together with sloughed-off root cap cells, instigates the rhizosphere effect (Pugh, 1980; Funck-Jensen & Hockenhull, 1984). Within a period of a few days the root surface becomes colonized by fungi which appear to grow along the root until they reach the next colonist. These root surface inhabitants out-compete the primary ruderal rhizosphere microflora, and the root surface moves to a situation of nutrient depletion (high stress/ low disturbance category) where the nutritionally least fit fungi have died out (Pugh, 1980).

Our study demonstrates that *G. roseum* has the ability to colonize young barley roots growing in sand and field soil, and thus we confirm the findings of Pugh (1980), that *G. roseum* is a root surface inhabitant. Pugh (1980) also reported that *G. roseum* is a soil competitor and it is of interest, that the fungus has been found to be a non-pathogenic, systemic colonizer of soybeans particularly of roots and the older parts of stems (Mueller & Sinclair, 1986). These authors concluded that colonization and secondary spread of *G. roseum* are either very slow or are favoured by the presence of senescent tissue.

Due to the contamination problem experienced when isolating *G. roseum* from roots grown in field soil, we are attempting to improve the selectivity of the medium used for dilution plating. Our group is also developing the use of isoenzymes and PCR-based methods for the detection and quantification of released microorganisms. It is expected that these methods will give us better possibilities to study long-term population dynamics in field experiments, and we plan to study root colonization of *G.*

roseum at harvest time and carry-over of inoculum to the following crop in field plots monocropped with cereals.

6. References

Baker, R. (1991) Induction of rhizosphere competence in the biocontrol fungus *Trichoderma* in Keister, D.L & Cregan, P.B. (eds.), *The Rhizosphere and Plant Growth*, Kluwer Acadademic Publishers, Dordrecht pp. 221-228..

Chao, W.L., Nelson, E.B., Harman, G.E. & Hoch, H.C. (1986) Colonization of the rhizosphere by biocontrol agents applied to seeds. *Phytopathology* **76**, 60-65.

Elad, Y. & Chet, I. (1983) Improved media for isolation of *Trichoderma* spp. or *Fusarium* spp. *Phytoparasitica* **11**, 55-58.

Funck-Jensen, D. & Hockenhull, J. (1984) Root exudation, rhizosphere microorganisms and disease control. *Växtskyddsnotiser* **48**, 49-54.

Knudsen, Inge M.B., Hockenhull, J & Jensen, D.F. (1995) Biocontrol of seedling diseases caused by *Fusarium culmorum* and *Bipolaris sorokiniana*: effects of selected fungal antagonists on growth and yield components. *Plant Pathology* **44**, 467-477.

Kraft, J.M. & Papavizas, G.C. (1983) Use of host resistance, *Trichoderma* and fungicides to control soilborne diseases and increase seed yields of pea. *Plant Disease* **67**, 1234-1237.

Krauss, U. & Deacon, J.W. (1994) Water-facilitated transport of a pimaricin-resistant strain of *Mucor hiemalis* in the rhizosphere of groundnut (*Arachis hypogaea* L.) in a Malawian ferric luvisol. *Soil Biology & Biochemistry* **26**, 977-985.

Madsen, E.L. & Alexander, M. (1982) Transport of *Rhizobium* and *Pseudomonas* through soil. *Soil Science Society of America Journal* **46**, 557-560.

Merriman, P.R., Price, R.D., Kollmorgen, J.F., Piggott, T. & Ridge, E.H. (1974) Effect of seed inoculation with *Bacillus subtilis* and *Streptomyces griseus* on the growth of cereals and carrots. *Australian Journal of Agricultural Research* **25**, 219-226.

Mueller, J.D. & Sinclair, J.B. (1986) Occurrence and role of *Gliocladium roseum* in field-grown soybeans in Illinois. *Transactions of the British Mycological Society* **86**, 677-680.

Peterson, E.A. (1959) Seed-borne fungi in relation to colonization of roots. *Canadian Journal of Microbiology* **5**, 579-582.

Pugh, G.J.F. (1980) Strategies in fungal ecology. *Transactions of the British Mycological Society* **75**, 1-14.

Rouatt, J.W. (1959) Initiation of the rhizosphere effect. *Canadian Journal of Microbiology* **5**, 67-71.

Stenton, H. (1958) Colonization of roots of *Pisum sativum* L. by fungi. *Transactions of the British Mycological Society* **41**, 74-80.

Windels, C.E. (1981) Growth of *Penicillium oxalicum* as a biological control seed treatment of pea seed in soil. *Phytopathology* **71**, 929-933.

USE OF A METALAXYL TOLERANT *PYTHIUM OLIGANDRUM* ISOLATE FOR SELECTIVELY FOLLOWING ITS POPULATION IN SOIL AND ITS COLONIZATION OF ROOTS OF DISEASED AND NON-DISEASED PLANTS

ANNE METTE MADSEN

Department of Plant Biology, The Royal Veterinary and Agricultural University, 40 Thorvaldsensvej, DK-1871 Frederiksberg C, Copenhagen, Denmark

Summary

When encysted zoospores of a metalaxyl-tolerant mutant of *Pythium oligandrum* were added to moist, unsterile soil at 20° C, 88% germinated within one hour. Most probable number analysis (MPN) showed that the fungus was recovered at one-half to one-third of the original level at 28 d, and it was still detected at 80 d. Growing cress plants did not affect the population, but decaying cress enabled the population to establish at a higher level.

The ability of *P. oligandrum* to colonize roots of diseased and non-diseased plants was compared. In the first study, the plants were 8 weeks old *Exacum affine* growing in peat, and the diseased plants were naturally infected by *Botrytis* sp. in the root collar. Four days after inoculation with *P. oligandrum* mycelium, the fungus was detected on 10.7% of the diseased root segments, and on 3.9% of the non-diseased root segments. Three weeks after inoculation, *P. oligandrum* had colonized 55.7% of the root segments of the diseased plants, and only 2.0% of the root segments of the non-diseased plants. In a second study, cress seedlings were placed in unsterile soil, and *P. oligandrum* was inoculated into the soil as encysted zoospores. The roots utilized were either colonized by *P. ultimum* or uninfected roots either attached (AR) or removed (RR) from the hypocotyl. After 64 hours, *P. oligandrum* had not colonized any of the roots at a detectable level. After 7 days, *P. oligandrum* was growing on 26.2%, 16.3% or 8.6% of the *P. ultimum* colonized, the RR or the AR segments respectively. This significant difference in the ability of *P. oligandrum* to colonize roots of diseased and non-diseased plants is discussed.

D. F. Jensen et al. (eds.), Monitoring Antagonistic Fungi Deliberately Released into the Environment, 39–48.
© *1996 Kluwer Academic Publishers. Printed in the Netherlands.*

1. Introduction

Pythium oligandrum Drechsler is an aggressive mycoparasite, which can penetrate host fungi within *ca* 5-15 min following contact (e.g. Deacon, 1976; Laing & Deacon, 1991). It is widely distributed in agricultural and horticultural soils (Plaats-Niterink, 1975; Mulligan & Deacon, 1992; Ribeiro & Butler, 1992; White, Wakeham & Petch, 1992). In experimental glasshouse conditions it has been shown to protect many crop plants against seedling diseases caused by *Pythium* spp. and other fungi (e.g. Vesely, 1979; Al-Hamdani, Lutchmeah & Cooke, 1983; Martin & Hancock, 1987; Walther & Gindrat, 1987). Oospores of *P. oligandrum* are often used as an inoculum in biocontrol experiments (e.g. Walther & Gindrat, 1987; McQuilken; Whipps & Cooke, 1992b), but one disadvantage of using oospores is their relatively poorly and slowly germination (Walther & Gindrat, 1987; McQuilken, Whipps & Cooke, 1992a). Alternative types of inoculum are zoospores, cysts, mycelia or zoosporangia. Cysts of *P. oligandrum* germinate fast *in vitro* and are easily produced (Madsen, Robinson & Deacon, 1995) as are mycelia and zoosporangia, however their behaviour in soil is poorly understood.

In natural infested soil *P. oligandrum* is sometimes isolated from roots colonized with other pythia (Drechsler, 1939; Middleton, 1952; Hendrix & Cambell, 1970; Pieczarka & Abawi, 1978; Trapero-Casas, Kaiser & Ingram, 1990; Cother & Cilbert, 1993), but there is no clear evidence of the ability of *P. oligandrum* to colonize roots. In one investigation *P. oligandrum* had colonized 19.2% of the sugar-beet root length after 3 days and only 8.9% after 5 days (Martin & Hancock, 1987). In another investigation the fungus could not colonize the cress and sugar-beet roots more than one cm below the *P. oligandrum* oospore coated seeds (McQuilken, Whipps & Cooke, 1990).

In this study a metalaxyl-tolerant mutant of *P. oligandrum* was used for selectively following the population over a period of time in an unsterile soil and the colonization of roots of both diseased and non-diseased plants. Mycelia or encysted zoospores were used as soil inoculum.

2. Material and methods

2.1 PRODUCTION OF A METALAXYL TOLERANT MUTANT OF *PYTHIUM OLIGANDRUM* AND PRODUCTION OF CYSTS

A metalaxyl tolerant mutant of *P. oligandrum* (isolate MM1) was produced by UV irradiation as described in Madsen *et al.* (1995). The isolate was used in all experiments present here. Cysts of the metalaxyl tolerant isolate were produced according to Madsen *et al.* (1995).

2.2 GERMINATION OF CYSTS AND RELEASE OF ZOOSPORES IN SOIL

Encysting zoospores were carefully concentrated on a cellulose nitrate filter (pore size 3.0 μm, Sartorius), and the filter was immediately placed on a soil sample (brought to full water-holding capacity) and covered with soil. After one hour most of the soil was removed, and the percentage of germinated cysts was estimated by microscopy (x400) after staining with 0.6 % anilineblue. One hundred cysts were recorded on each of the 3 replicates.

Zoosporangia were produced in a liquid medium, washed in 4°C cold water, and transferred to a supersaturated unsterile soil. After 4 h at 20°C, samples were collected for microscopy.

2.3 SURVIVAL OF CYSTS IN SOIL

The soil was prepared as described in Madsen *et al.* (1995). In one experiment, with 25 g soil per dish and 4 replicates, the soil was brought to full water-holding capacity and contained either 88 or 755 cysts of *P. oligandrum* g^{-1} moist soil (equivalent to 148 or 1260 cysts g^{-1} air-dry soil). In half of the replicates with high cyst concentration, cress was sown after 8 days. In a second experiment, with 40g soil per dish and 8 replicates, the soil was brought to 50% water-holding capacity with 1690 cysts g^{-1} moist soil (2260 g^{-1} air-dry soil). The dishes were incubated at *ca* 20°C, with Parafilm seals to prevent moisture loss.

P. oligandrum was detected on sectors of agar with 50 μg metalaxyl ml $^{-1}$ precolonized by *Fusarium culmorum* as described in Madsen *et al.* (1995). The numbers of replicate 'positive' sectors at different soil dilutions were used to calculate the most probable number (MPN) of infective units of *P. oligandrum* in the undiluted soil, using tables in Fisher & Yates (1963).

2.4 INOCULUM FOR ROOT COLONIZATION

In the first experiment mycelial mats from 3-days old 10% V8 juice cultures of *P. oligandrum* (MM1) were homogenized by an ultra-turrax for 2 min. at 13500 rpm. Twenty-eight *Exacum affine* plants were inoculated with 8 ml *P. oligandrum* suspension (approx. 3.0 mg dried mycelium) per pot under aseptic conditions.

In a second experiment with cress (*Lepidium sativum*) roots, encysted zoospores were used as inoculum in the concentration 1.5 x 10⁴ cysts g^{-1} dry soil.

2.5 ROOT COLONIZATION

In the first experiment, twenty-eight 8-weeks old *E. affine* plants were utilised. Fourteen of the plants were non-diseased, the other fourteen plants were infected by *Botrytis* sp. in the root collar. The roots of both groups of plants were still white, but the roots of the diseased plants were a little thinner.

The experiment was performed in 10 cm-diameter round pots with *ca* 45 g peat (no. 2, Steen Røgild, pH 6) per pot, at 20° C. Plants were watered daily but not fertilized.

The roots of 4 and 10 plants were collected both on 4 days and 21 days following inoculation with *P. oligandrum*. Roots were rinsed and cut into segments of 1 cm. The number of *P. oligandrum* colonized root segments was estimated. Some root segments were used directly for microscopy, and further 72 root segments from each plant were placed on sections of a *Pythium* selective medium (Jeffers & Martin, 1986) with an addition of 40 µg metalaxyl ml^{-1}.

In the second experiment the colonization of cress roots was examined. Cress seeds were germinated on 1% water-agar, afterwards 30 plants were placed on Potato Dextrose- agar colonies of *P. ultimum* for 24 h at 20° C to allow colonization of the roots. The plants were moved to dishes with 15 g of soil, and the roots were covered with another 15 g of soil. The roots of another 30 plants were cut off the hypocotyl (RR), and also placed in the soil. Finally, 30 untreated cress plants (AR) were placed carefully in the soil. The soil (a Danish agricultural soil pH 6.4, sandy loam) was prepared as in Madsen *et al.* (1995) and saturated with a suspension of encysted zoospores of *P. oligandrum*.

Twenty and ten roots from each treatment were collected after 64 h and 7 days respectively, and the colonization by *P. oligandrum* was investigated as for the *E. affine* roots.

3. Results

3.1 GERMINATION OF *P. OLIGANDRUM* CYSTS AND RELEASE OF ZOOSPORES IN SOIL

When cysts were inoculated within an unsterile soil, 88% germinated within one hour, and when the cysts were placed in water 91% germinated within the same period.

Four h after zoosporangia were placed in the soil zoospores were detected. Likewise when zoospores are produced on agar in sterile water. Using microscopy many of the sporangia in the soil were also observed to be empty after 4 h.

3.2 SURVIVAL OF *P. OLIGANDRUM* CYSTS IN SOIL

In the first experiment, soil was incubated at full water-holding capacity with 755 or 88 cysts g^{-1} moist soil. With the higher cyst concentration, *P. oligandrum* was detected on nearly all (22 or 23) of the 24 replicate agar sectors that received undiluted soil at samplings made after 4h, 24h, 3 days and 8 days. The detection frequency then declined to 16 out of 24 replicate platings at 13 days, 23 out of 36 (20 days), 7 out of 36 (42 days) and 5 out of 36 (80 days). At the lower cyst concentration *P. oligandrum* was detected on all platings made after 24h but the frequency of detection then declined to only 4 out of 24 platings at 8 days, 2 out of 36 (20 days) and none out of 36 (42 days). This experiment also involved a treatment in which cress seeds were sown in the soil dishes at 8 days; 5 and 12 days after cress was sown there was no difference in the frequency of detection of *P. oligandrum*. At day 42 and 80 the cress plats were decaying, and the detection frequency was higher in these soils. *P. oligandrum* could be detected in 20 out of 36 sections (day 42) and 22 out of 36 sections (day 80) if cress was sown in the soil but only in 7 out of 36 (day 42) and in 5 out of 36 (day 80) in the soil with no amendment.

In the second experiment (soil at 50% of water-holding capacity, with 1690 cysts g^{-1} moist soil), the most probable number of 'infective units' g^{-1} moist soil declined with time (Table 1), but in every soil dish the introduced strain of *P. oligandrum* was still detectable after 28 days (the longest tested). The means of the replicates (Table 1) show a clear trend: the estimated population changed little over the first 5 days but declined to one-half to one-third of the original by day 14 and then remained stable until at least 28 days (Madsen *et al.*, 1995).

3.3 COLONIZATION OF ROOTS IN SOIL AND PEAT

All the *E. affine* plants (without *Botrytis* sp.) remained healthy following 3 weeks growth with *P. oligandrum*. *P. oligandrum* was detected by microscopy on some roots and root collars as oospores, and it had survived in the peat of both the diseased and non-diseased plants.

When the *E. affine* roots were collected 4 days after inoculation with *P. oligandrum*, the fungus was detected on 10.7% of the diseased roots and on 3.9% of the non-diseased roots (Table 2). When the roots were collected after 3 weeks, *P. oligandrum* was detected on 415 out of 1440 one cm long root segments. *P. oligandrum* was detected on root segments on all the 10 diseased plants, but only on root segments of 4 of the actively growing plants.

Cress roots were collected at 64 hours and 7 days following addition of *P. oligandrum* cysts to the soil. *P. oligandrum* did not grow out from neither *P. ultimum* colonized roots, RRs nor ARs collected after 64 hours. *P. ultimum*, however, did grow out from the precolonized roots. When the roots were collected after 7 days, *P. oligandrum* was growing out of 26.2%, 16.3% or 8.6% of the *P. ultimum* colonized, the RR or the AR segments, respectively.

Table 1. Estimated numbers of infective units of *P. oligandrum* g⁻¹ moist soil at different times after cysts (1690 g⁻¹ moist soil) were added to soil at 50% water-holding capacity.

Replicate dish of soil	Time after addition of cysts to soil				
	4h	120h	14 days	21 days	28 days
1	102	65	14	17	8
2	65	156	13	24	17
3	31	65	17	26	24
4	31	27	17	11	11
5	39	27	7	13	16
6	27	51	17	26	8
7	27	51	17	26	16
8	35	51	31	11	24
Mean	45	44	17	20	16

(Madsen *et al.*, 1995)

Table 2. Percentage of diseased and non-diseased *E. affine* root segments colonized by *P. oligandrum* at day 4 and 21.

Day	Percentage colonized root segments	
	Diseased (%)	Non-diseased (%)
4	10.7 ± 7.9	3.9 ± 5.6
21	55.7 ± 10.4	2.0 ± 1.1

4. Discussion

The use of a metalaxyl tolerant mutant of *P. oligandrum* made it possible to selectively follow the population of *P. oligandrum* cysts over a period of time, and to investigate the colonization of roots by *P. oligandrum* in unsterile soils or peat. The metalaxyl-tolerant strain of *P. oligandrum* could be detected at least 80 days after cysts were inoculated to soil. Most-probable number analysis (Table 1) suggests that the population of *P. oligandrum*, from cysts introduced into soil, declined to one-half to one-third of the original (4h sampling) after 28 days when no organic-material was added. Moreover, most of this decline seemed to occur between 5 and 14 days, and the residual population then stabilised. Mulligan, Jones & Deacon (1995) also found that *P. oligandrum* stabilised at a certain level, when oospores were added to the soil. When cress plants were decaying in the soil, the population did not decrease as in the control, and this is most likely related to the nutritional value of the plant debris. In other studies (Mulligan *et al.*, 1995) it was also possible to manipulate the *P. oligandrum* population by adding appropriate organic-material to a soil. Different investigations also suggest that *P. oligandrum* is most common in soils with high content of organic material (Foley & Deacon, 1985; Mulligan & Deacon, 1992) as other *Pythium* spp. are more common in arable soils than in undisturbed sites (e. g. Plaats-Niterink, 1975; Foley & Deacon, 1985; Pankhurst, McDonald & Hawke, 1995).

Using microscopy *P. oligandrum* oogonia were observed on roots of diseased *E. affine* plants. By placing the roots on a *Pythium* selective media supplemented with metalaxyl (40 μgml^{-1}), it was shown that it was the introduced isolate that had colonized the roots. Thus it was possible for the mycelia and cysts to establish in the peat and to colonize the diseased roots. Cysts of *P. oligandrum* germinated rapidly in soil and in water without any trigger. In another study encysted zoospores of *P. aquatile* are indirectly shown to germinate in soil, since the retrievability of cysts inoculated into soil increased over time (Hardman, Pike & Dick, 1989). In contrast Luna & Hine (1964) found that encysted zoospores of *P. aphanidermatum* in both sterile and non-sterile soil only germinated upon the addition of exogenous nutrients to the soil. So, it is not a general feature that cysts have a high germination rate in soil. Under optimum laboratory conditions only up to 30% *P. oligandrum* oospores germinate within 12 h (McQuilken *et al.*, 1992a), while 88% cysts germinate in soil. It is therefore tempting to try encysted zoospores or sporangia as inoculum for biocontrol.

The population of *P. oligandrum* in soil was unaffected by actively growing cress plants, and furthermore it was shown that *P. oligandrum* only had a good ability to colonize *E. affine* roots of diseased plants. *P. oligandrum* is often reported to be isolated from roots with plant pathogens (Drechsler, 1939; Middleton, 1952; Hendrix & Cambell, 1970; Pieczarka & Abawi, 1978; Trapero-Casas *et al.*, 1990; Cother & Cilbert, 1993), but this great difference between its ability to colonize diseased and non-diseased roots (Table 2) has not been shown before. In other experiments it was shown that *P. oligandrum* had a significantly greater ability to colonize uncolonized wheat flag leaves, than precolonized leaves (Mulligan *et al.* 1995), and Martin & Hancock (1986) showed that *P. oligandrum* seldom colonized the same leaf debris as *P. ultimum*, although both of the fungi had an individual high frequency of colonization. It can seem contradictory, that *P. oligandrum* on one hand is frequently isolated from roots with plant pathogens, it has a good ability to colonize precolonized agar (Deacon, 1976; Foley & Deacon, 1986); and it is more often isolated from diseased than from non-diseased roots, and on the other hand, that *P. oligandrum* seldom colonizes leaf debris colonized by *P. ultimum* or other fungi; and that the precolonized roots have to be old before *P. oligandrum* can be detected on them. It is not yet possible to explain these contradictions, but some of them may possibly be explained by a) the resistance of the growing plant to the mycoparasite, b) the resistance of the precolonizing fungi to the mycoparasite or c) the occupation by different fungal species of different niches on the roots. Related to a) *P. oligandrum* is not able to utilize cellulose (Tribe, 1966; Deacon, 1979) and it does not produce extracellular cellulases when it is growing on cell walls of *P. ultimum* (Madsen, unpublished results), even though it can coil around *P. ultimum* hyphae and penetrate them (Berry, Jones & Deacon, 1993). Related to b) preliminary results (Madsen, unpublished data) show that older *P. ultimum* hyphae are more susceptible to *P. oligandrum* than younger hyphae. Furthermore *P. oligandrum* grows faster on old *P. ultimum* cultures than on young cultures (Deacon, 1976). Finally *P. oligandrum* was more often detected on roots of *E. affine* plants attacked by *Botrytis* sp. than on cress roots attacked by *P. ultimum*, and this may possibly be related to the results of Foley & Deacon (1986) and Laing & Deacon (1991) who found that *Botrytis* sp. is a more susceptible host than *P. ultimum*. Related to c) it is shown that zoospores of *Pythium* and *Phythophthora* species accumulate at specific sites of roots (reviewed by Deacon & Donaldson, 1993), in the same way different species of fungi may be adapted to colonize different zones on the roots.

P. oligandrum is known to be an aggressive mycoparasite, and its ability to colonize organic material from decaying plants can be expected to be an advantage in using the fungi in biological control. Some plant roots like barley and wheat are shown to have layers of dead cortex (Henry & Deacon, 1981), and these layers may help *P. oligandrum* to establish on the roots and thereby protect them. The results that *P. oligandrum* is not a good colonizer of non-diseased roots, are consistent with the results of McQuilken *et al.* (1990). According to this, one alternative suggestion of using *P. oligandrum* in biocontrol, is to use it together with a non-pathogen rhizosphere competent host, for example *Phialophora radicicola* var *graminicola*, as it is a widespread (Deacon, 1973a,b; Skipp & Christensen, 1989), very susceptible non pathogen host of *P. oligandrum* (Deacon, 1976) which colonizes grass and cereal root cells (Deacon, 1980; Henry & Deacon, 1981, Skipp & Christensen, 1989).

Acknowledgements
The population study of zoospore cysts in soil I did at Edinburgh University, and I am very pleased to thank Dr. J. W. Deacon for supervision, and the Danish Agricultural and Veterinary Research Council for financial support. I also want to thank Mette Lübeck and Helen Lynne Robinson for their aid.

REFERENCES

Al-Hamdani, A.M., Lutchmeah, R. S. & Cooke, R.C. (1983). Biological control of *Pythium ultimum* induced damping-off by treating cress seeds with the mycoparasite *Pythium oligandrum*. *Plant Pathology* **32**, 449-454.

Berry, L. A., Jones, E. E. & Deacon, J. W. (1993). Interaction of the mycoparasite *Pythium oligandrum* with other *Pythium* species. *Biocontrol Science and Technology* 3, 247-260.

Cother, E. J. & Gilbert, R. L. (1993). Comparative pathogenicity of *Pythium* species associated with poor seedling establishment of rice in southern Australia. *Plant Pathology* 42, 151-157.

Deacon, J. W. (1973a). *Phialophora radicicola* and *Gaeumannomyces graminis* on roots of grasses and cereals. *Transactions of the British Mycological Society* 61, 471-485.

Deacon, J. W. (1973b). Factors affecting occurrence of the *Ophiobolus* patch disease of turf and its control by *Phialophora radicicola*.*Plant pathology* 22, 149-155.

Deacon, J. W. (1976). Studies on *Pythium oligandrum*, an aggressive parasite of other fungi. *Transactions of the British Mycological Society* 66, 383-391.

Deacon, J. W. (1979). Cellulose decomposition by *Pythium* and its relevance to substrat-groups of fungi. *Transactions of the British Mycological Society* 72, 469-477.

Deacon, J. W. (1980). Ectotrophic growth by *Phialophora radicicola* var. *graminicola* and other parasites of cereal and grass roots. *Transactions of the British Mycological Society* 75, 158-160.

Deacon, J. W. & Donaldson, S. P. (1993). Molecular recognition in the homing sequence of zoosporic fungi, with special reference to *Pythium* and *Phytophthora*. *Mycological Research* 97, 1153-1171.

Drechsler, C. (1939). Several species of *Pythium* causing blossom-end rot of watermelons. *Phytopathology* 29, 391-422.

Fisher, R. A. & Yates, F. (1963). *Statistical Tables for Biological, Agricultural and Medical Research.* Oliver & Boyd, Edinburgh.

Foley M. F. & Deacon J. W. (1985). Isolation of *Pythium oligandrum* and other necrotrophic mycoparasites from soil. *Trans. Br. mycol. Soc.* 85, 631-639.

Foley M. F. & Deacon J. W. (1986). Susceptibility of *Pythium* spp. and other fungi to antagonism by the mycoparasite *Pythium oligandrum*. *Soil Biology & Biochemistry* 18, 91-95.

Hardman, J. M., Pike, D. J. & Dick, M. W. (1989). Short-term retrievability of *Pythium* propagules in simulated soil environments. *Mycological Research* 93, 199-207.

Hendrix, J. W. & Cambell W. A. (1970). Distribution of *Phytophthora* and *Pythium* species in soils in the Continental united states. *Canadian Journal of Botany* 48, 377-384.

Henry, C. M. & Deacon J. W. (1981). Natural (non-pathogenic) death of the cortex of wheat and barley seminal roots, as evidenced by nuclear staining with acridine orange. *Plant and Soil* 60, 225-274.

Jeffers, S. N. & Martin, S. B. (1986). Comparison of two media selective for *Phythophthora* and *Pythium* species. *Plant Disease* 70, 1038-1043.

Laing, S. A. K. & Deacon, J. W. (1991). Video microscopical comparison of mycoparasitism by *Pythium oligandrum*, *P. nunn* and an unnamed *Pythium* species. *Mycological Research* 95, 469-479.

Luna, L. V. & Hine, R. B. (1964). Factors influencing saprophytic growth of *Pythium aphanidermatum* in soil. *Phytopathology* 54, 955-959.

Madsen, A. M., Robinson, H. L. & Deacon, J. W. (1995) (in press). Behaviour of zoospore cysts of the mycoparasite *Pythium oligandrum* in relation to biocontrol of plant pathogens. *Mycological Research*.

Martin, F. N. & Hancock, J. G. (1986). Association of chemical and biological factors in soils suppressive to *Pythium ultimum*. *Phytopathology*, 76, 1221-1231.

Martin, F. M. & Hancock, J. G. (1987). The use of *Pythium oligandrum* for biological control of pre-emergence damping-off caused by *Pythium ultimum*. *Phytopathology* 77, 1013-1020.

McQuilken, M. P., Whipps, J. M. & Cooke, R. C. (1990). Control of damping-off in cress and sugar-beet by commercial seed-coating with *Pythium oligandrum*. *Plant Pathology* 39, 452-462.

McQuilken, M. P., Whipps, J. M. & Cooke, R. C. (1992a). Effects of osmotic and matric potential on growth and oospore germination of the biocontrol agent *Pythium oligandrum*. *Mycological Research* 96, 588-591.

McQuilken, M. P., Whipps, J. M. & Cooke, R. C. (1992b). Use of oospore formulations of *Pythium oligandrum* for biological control of *Pythium* damping-off in cress. *Journal of Phytopathology* 135, 125-134.

Middelton J. T. (1952). *Pythium* seed decay and seedling blight of *Pisum sativum*. *Phytopathology* 42, 516.

Mulligan, D. F. C. & Deacon J. W. (1992). Detection of presumptive mycoparasites in soil placed on host-colonized agar plates. *Mycological Research* 96, 605-608.

Mulligan, D. F. C., Jones, E. E. & Deacon, J. W. (1995) (in press). Monitoring and manipulation of *Pythium*

48

oligandrum, P. mycoparasiticum and *Papulspora* species in soil. *Soil Biology & Biochemistry* **27**, 1333-1343.

Pankhurst, C. E., McDonald, H. J. & Hawke, B. G. (1995). Influence of tillage and crop rotation on the epidemiology of *Pythium* infections of wheat in a red-brown earth of south Australia. *Soil Biol. Biochem.* **27**, 1065-1073.

Pieczarka, D. J. & Abawi, G. S. (1978). Populations and biology of *Pythium* species associated with snap bean roots and soils in New York. *Phytopathology* **68**, 409-416.

Plaats-Niterink, A.J. van der (1975). Species of *Pythium* in the Netherlands. *Netherlands Journal of Plant Pathology* **81**, 23-37.

Ribeiro, W. R. C. & Butler, E. E. (1992). Isolation of mycoparasitic species of *Puthium* with spiny oogonia from soil in California. *Mycological Research* **95**, 857-862.

Skipp, R. A. & Christensen, M. J. (1989). Fungi invading roots of perennial ryegrass (*Lolium perenne* L.) in pasture. *New Zealand Journal of Agricultural Research* **32**, 423-431.

Trapero-Casas, A., Kaiser, W. J. & Ingram, D. (1990). Control of *Pythium* seed rot and preemergence damping-off of chickpea in the U. S. pacific NorthWest and Spain. *Plant Disease* **74**, 563-569.

Tribe, H. T. (1966). Interaction of soil fungi on cellulose film. *Transactions of the British Mycological Society* **49**, 457-466.

Vesely, D. (1979). Use of *Pythium oligandrum* to protect emerging sugar-beet. In *Soil-borne Plant Pathogens* (ed. Schippers, B & Gams, W.), pp. 593-595. Academic Press, London.

Walther, D. & Gindrat, D. (1987). Biological control of *Phoma* and *Pythium* damping-off of sugar-beet with *Pythium oligandrum*. *Journal of Phytopathology* **119**, 167-174.

White, J. G., Wakeham, A. J. & Petch, G. M. (1992). Deleterius effect of soil-applied metalaxyl and mancozeb on the mycoparasite *Pythium oligandrum*. *Biocontrol Science and Technology* **2**, 335-340.

RECOVERY AND DETECTION OF DEUTEROMYCETE CONIDIA FROM SOIL:

Examplified by isolation of Penicillium verrucosum *Dierckx on a selective and diagnostic agar medium*

SUSANNE ELMHOLT AND HELLE HESTBJERG
Danish Institute of Plant and Soil Science, Department of Soil Science, Research Centre Foulum, P.O. Box 23, DK-8830 Tjele

1. Abstract

The field ecology of *Penicillium verrucosum* (important Ochratoxin A producing storage fungus) is poorly understood. It is for example not known whether the fungus can survive, proliferate, and infect cereal plants in the field. For such studies, a sensitive method is needed to screen the presence of the fungus in naturally and artificially infested soils. Prior to experiments with deliberate release of *P. verrucosum* into the field environment, preliminary experiments were performed regarding four different media (V8, YES, DRYES, and DYSG). Only DYSG has selective and indicative properties that makes it suitable as a medium for recover and easy detection of *P. verrucosum* from a population of deuteromycete soil fungi. The recovery and detection level of *P. verrucosum* on DYSG was determined on Petri dishes, plated with mixtures from a suspension of a non-spiked soil and decreasing amounts of a suspension from a soil, spiked with conidia of *P. verrucosum*. Results showed a very high recovery (91% and 112%) of *P. verrucosum* and a detection level of 300 *P. verrucosum* CFU/g dry soil. In conclusion: given a good selective medium, classical microbiological methods may well be used for monitoring fungi deliberately released into the environment.

2. Introduction

Penicillium verrucosum (chemotype II) is important to grain deterioration in temperate regions, including Denmark. It can produce the mycotoxins Ochratoxin A (OA) and citrinin (Frisvad & Filtenborg, 1989). OA has carcinogenic, teratogenic, and immunotoxic properties, which have initiated numerous studies on grain contamination and toxicology in relation to man and animals (e.g. Hult *et al.*, 1982; Büchmann & Hald, 1985; Breitholz-Emanuelson *et al.*, 1993; Madhyastha *et al.*, 1993).

D. F. Jensen et al. (eds.), Monitoring Antagonistic Fungi Deliberately Released into the Environment, 49–55.
© *1996 Kluwer Academic Publishers. Printed in the Netherlands.*

Many mycotoxin producing fungi, including *P. verrucosum*, are regarded as storage fungi and their occurrence and behaviour have been studied almost exclusively in storage conditions. Nevertheless many experts on mycotoxins are convinced that the primeval habitat of these fungi is soil organic material. The ecology of *P. verrucosum* is currently being studied at the Department of Soil Science. The purpose is to elucidate whether the fungus can survive, proliferate, and infect cereal plants in the field.

If present, *P. verrucosum* probably occurs in low numbers in arable soils. A sensitive, rapid, and cheap method is needed to screen its presence in naturally as well as artificially infested soils. In this respect, the methodology is comparable to that of introduction and re-isolation of antagonists from soil.

P. verrucosum grows willingly on most agar media. A number of attempts have been made to develop media that are both selective and indicative (diagnostic) of the species for use in mycological analyses of food and feed products.

Yeast Extract Sucrose (YES) agar was originally recommended as excellent for the induction of mycotoxin formation by numerous food-borne fungi. It is also diagnostic of *P. verrucosum*, because the fungus produces a characteristic brownish red non-diffusible pigment on the agar reverse (Frisvad, 1983). The selective principle of YES is a lowered water activity which allows growth of moderately xerotolerant fungi and at the same time partially inhibits fast growing *Zygomycetes*, like *Mucor*, *Rhizopus*, and *Absidia*.

The selectivity of YES against fast growing fungi has been improved by addition of pentachlornitrobenzene + rose bengal (PRYES agar) or 2,6-dichloro-4-nitroaniline (=botran, =dichloran) + rose bengal (DRYES agar) (Frisvad, 1983; Frisvad *et al.*, 1986).

Recently, a selective and diagnostic medium has been recommended for detection of *P. verrucosum* in cereal products, bread and meat, i.e. Dichloran Yeast extract Sucrose 18% Glycerol (DYSG) agar (Frisvad *et al.*, 1992). It has been developed from YES and its selectivity is based on dichloran (as in DRYES) and a very low water activity, which is achieved by means of 220 g/l glycerol.

Prior to experiments with deliberate release of *P. verrucosum* into the field environment, abundance, recovery, and detection level for conidia in soil, was studied on four different media with the main emphasis on DYSG.

3. Materials and methods

3.1. SOIL AND FUNGUS

Soil I was sampled at Foulumgård (56°30'N, 9°34'E) in a mixed ley of grass and clover to a depth of 10 cm on May 5th. Soil II, used for spiking, was sampled in a wheat field. *P. verrucosum*, IBT 5075, isolated from Danish grown barley, was used for the experiment (Lund *et al.*, 1992). Twenty Petri dishes with malt extract agar (MEA) (Pitt, 1979) were inoculated at three points and incubated at 25°C for 10 days. Forty ml Na-peptone dilution medium were used to wash off conidia from the plates and the suspension was mixed thoroughly into 200 g fresh, sieved (<2 mm) Soil II.

3.2. EXPERIMENTAL DESIGN

Using plate dilution, the number of CFU/g dry soil was determined for the non-spiked soil (Soil I) and the spiked soil (Soil II) on 4 different media: V8-juice agar (Gams *et al.*, 1987), YES (Frisvad & Filtenborg, 1989), DRYES (Frisvad, 1983), and DYSG (Frisvad *et al.*, 1992). V8-agar was used as a reference medium to assure that Soil I represented a typical arable soil microflora. From Soil I, soil equivalent to 8.2 g dry soil was suspended into 41.8 g water (dilution 1.6×10^{-1}). 0.1 ml of dilutions 1.6×10^{-1} to 1.6×10^{-5} was plated onto two Petri dishes of each of the four media. From Soil II, soil equivalent to 5 g dry soil was suspended into 45 ml water (dilution 1.0×10^{-1}). 0.1 ml of dilutions 1.0×10^{-1} to 1.0×10^{-6} was plated onto two Petri dishes of each of the four media. For the recovery and detection level experiment, aliquots of the spiked and non-spiked dilutes from Soil I and Soil II were placed on two replicate Petri dishes of DYSG according to the following design and mixed during spreading with a drigalsky spatula.

Experiment I:
Mix 1: 0.1 ml Soil I (dilution 1.6×10^{-1}) + 0.1 ml Soil II (dilution 1.0×10^{-2})
Mix 2: 0.1 ml Soil I (dilution 1.6×10^{-1}) + 0.1 ml Soil II (dilution 1.0×10^{-3})
Mix 3: 0.1 ml Soil I (dilution 1.6×10^{-1}) + 0.1 ml Soil II (dilution 1.0×10^{-4})
Mix 4: 0.1 ml Soil I (dilution 1.6×10^{-1}) + 0.1 ml Soil II (dilution 1.0×10^{-5})
Mix 5: 0.1 ml Soil I (dilution 1.6×10^{-1}) + 0.1 ml Soil II (dilution 1.0×10^{-6})

Experiment II:
Mix 6: 0.1 ml Soil I (dilution 1.6×10^{-2}) + 0.1 ml Soil II (dilution 1.0×10^{-3})
Mix 7: 0.1 ml Soil I (dilution 1.6×10^{-2}) + 0.1 ml Soil II (dilution 1.0×10^{-4})
Mix 8: 0.1 ml Soil I (dilution 1.6×10^{-2}) + 0.1 ml Soil II (dilution 1.0×10^{-5})
Mix 9: 0.1 ml Soil I (dilution 1.6×10^{-2}) + 0.1 ml Soil II (dilution 1.0×10^{-6})

The total number of CFU was determined after 8-9 days at 25°C (incubation in the dark), based on dilution 1.6×10^{-3} (Soil I) and dilution 1.0×10^{-5} (Soil II). Recovery and detection level of *P. verrucosum* were determined, using results from the soil suspension mixtures on DYSG. The number of *P. verrucosum* was determined by counting colonies with the characteristic "Terra-cotta coloured" reverse pigmentation on DYSG (D7-D8, according to Kornerup and Wanscher (1969)).

4. Results and discussion

4.1. PROPERTIES OF THE MEDIA

Table 1 shows the number of colony forming units (CFU) in Soil I and II on four different agar media. In the non-spiked Soil I, the results represent a mixed population of soil fungi. The CFU on DRYES is 3.9×10^5. Compared to DRYES, values were reduced by 22% (V8), 25% (YES), and 63% (DYSG), showing a very good selectivity of DYSG. In the spiked Soil II, all CFU represent *P. verrucosum*, which totally outnumbered naturally occurring fungi on the Petri dishes at all dilution levels. The CFU on DRYES is 5.95×10^7 and compared to this, values were reduced by 14%

(V8), 9% (YES), and 16% (DYSG). The results for Soil II confirm that *P. verrucosum* grows willingly on all tested media.

TABLE 1. Number of fungi (CFU/g oven-dry soil) in Soil I (non-spiked) and Soil II (*P. verrucosum*-spiked) on four different media (V8-agar, YES, DRYES, and DYSG).

Substrate	V8-agar	YES	DRYES	DYS18G
Soil	Colony forming units (CFU) / g oven-dry soil			
SOIL I	3.05×10^5	2.93×10^5	3.90×10^5	1.46×10^5
SOIL II	5.10×10^7	5.40×10^7	5.95×10^7	5.00×10^7

YES is a good indicative medium for *P. verrucosum* due to its characteristic reverse colour. However, its poor selectivity towards soil *Zygomycetes* (*Mucor*, *Mortierella*, and *Absidia*) makes a detection of *P. verrucosum* in low numbers in soil very unlikely.

DRYES was designed for detecting *P. verrucosum* in food products. In spite of its well-documented selective properties, the highest number of CFU in the present study was actually found on DRYES. Bearing in mind that differences in CFU results on the four media could not be confirmed statistically in these preliminary experiments, the lower number on V8 than on DRYES may be due to a masking on V8 of other fungi by fast growing *Zygomycetes*. In Soil II it may be due to the growth restricting properties of rose bengal allowing a higher number of developing colonies of *P. verrucosum* on DRYES than on V8. The violet brown reverse coulour of DRYES is indicative of the species in a mixed population of food-borne fungi but not in a mixed population of soil fungi. This is partly so because some of the *Penicillium* species, most commonly encountered in Danish arable soils, produce almost the same reverse colour, among them *P. canescens* and related species most of which belong to Subgenus *Furcatum* Sect. *Divaricatum*.

DYSG has selective and indicative properties, which were shown to be excellent for detection of *P. verrucosum* in a mixed soil population. The selectivity of the medium reduced the number of isolated fungi in Soil I by more than 60% (Table 1) due to the water activity reduction by glycerol. This included a near to complete elimination of *Zygomycetes* at all dilution levels. Due to the exclusion of rose bengal from DYSG, the indicative principle is much better and the reverse colour not as likely confused with commonly occurring soil fungi as on DRYES.

4.2. RECOVERY AND DETECTION

Table 2 shows calculated estimates on DYSG for the mixed suspensions of soil fungi (originating from the Soil I suspension) and of *P. verrucosum* (originating from the Soil II suspension). These estimates were based on the CFU results on DYSG in Table 1. In Experiment I, approximately 2500 viable propagules of soil fungi on DYSG were mixed with 5 to 50,000 propagules of *P. verrucosum* (ratio 500:1 to 1:20). In

Experiment II, approximately 250 viable propagules of soil fungi on DYSG were mixed with 5 to 5000 propagules of *P. verrucosum* (ratio 50:1 to 1:20).

Recovery: *P. verrucosum* was recorded on two replicate Petri dishes from each soil suspension mixture by means of its reverse pigmentation. In Mixtures 1, 2, 3, 6, and 7, the number of *P. verrucosum* was too high to count and accordingly, recovery could not be determined. In Mix 4, recovery was 91% and in Mix 8, recovery was 112%. The reason for the lower recovery in Mix 4 is probably a higher level of competition from the soil fungi, of which there are ten times more in Mix 4 than in Mix 8. To draw statistically sound conclusions from these comparisons, however, more Petri dish replications would be required. In Mixtures 5 and 9, recovery was not calculated, because colony numbers are so small that just one colony will affect recovery by 20%.

TABLE 2. Recovery (%) on DYSG of *Penicillium verrucosum* from mixed populations of soil fungi (detected colonies in relation to the estimated number).

EXPERIMENT I	Mix 1	Mix 2	Mix 3	Mix 4	Mix 5
Soil fungi estimated number on Petri dish	2394	2394	2394	2394	2394
P. verrucosum estimated number on Petri dish	50000	5000	500	50	5
P. verrucosum detected number on Petri dishes[**]	NC	NC	NC	47/44	2/4
Average recovery (%)	-	-	-	91	-[*]

EXPERIMENT II	Mix 6	Mix 7	Mix 8	Mix 9
Soil fungi estimated number on Petri dish	239	239	239	239
P. verrucosum estimated number on Petri dish	5000	500	50	5
P. verrucosum detected number on Petri dishes[**]	NC	NC	56/56	5/7
Average recovery (%)	-	-	112	-[*]

[*]   Numbers too small to calculate recovery.
[**]   Results from two replicate Petri dishes.
[***]   NC = not counted

<u>Detection level</u>: Mixtures 4, 5, and 9 illustrate the capacity of DYSG to discriminate low numbers of *P. verrucosum* in a mixed soil population. Mix 4 and 9 simulate a detection of 2% *P. verrucosum* in relation to soil fungi that are viable on DYSG. Mix 5 simulates 0.2 % of *P. verrucosum* in relation to soil fungi. This means that in the mixed Soil I population of fungi (1.46×10^5 CFU/g on DYSG), it would have been possible to detect *P. verrucosum* at about 300 CFU/g oven-dry soil! This number is much lower than for a general non-diagnostic medium, like V8. Using V8, a fungal isolate has to be re-isolated for identification. For cropped arable soils, this is normally done from 10^{-4} dilutions, which means that the fungus must be present in a concentration of minimum 10.000 CFU/g oven-dry soil.

5. Conclusion

The medium DYSG is excellent for detection of *P. verrucosum* in mixed populations of soil fungi! Due to its low water activity, the medium favours the xerotolerant properties of the species and at the same time excludes a major part of the soil fungi, being non-xerotolerant. The indicative principle facilitates diagnosis of very low numbers of *P. verrucosum* propagules. Recovery of *P. verrucosum* from spiked soil was very high (91% and 112%). The present experiments showed a detection level in mixed suspensions from spiked and non-spiked soils of 300 CFU/g dry soil. This can stand comparison with the detection level of many modern serological and DNA-based techniques. Classical microbiological methods may therefore well be used for monitoring fungi deliberately released into the environment, provided a good selective medium is available.

6. References

Breitholz-Emanuelson, A., Olsen, M., Oskarsson, A., and Hult, K. (1993). Ochratoxin A in Cow's Milk and in Human Milk with Corresponding Human Blood Samples, *J. AOAC Int.* **76**, 842-846.

Büchmann, N.B. and Hald, B. (1985). Analysis, occurrence and control of Ochratoxin A residues in Danish pig kidneys, *Food Additives and Contaminants* **2**, 193-199.

Frisvad, J.C. (1983). A selective and indicative medium for groups of *Penicillium viridicatum* producing different mycotoxins in cereals, *Journal of Applied Bacteriology* **54**, 409-416.

Frisvad, J.C., Thrane, U., and Filtenborg, O. (1986). Comparison of five media for enumeration of mesophilic mycoflora of foods, in A.D. King, J.I. Pitt, L.R. Beuchat, and J.E.L. Corry (eds.), *Methods for the mycological examination of food*, Plenum Press, New York, pp. 138-142. .

Frisvad, J.C. and Filtenborg, O. (1989) Terverticillate Penicillia: Chemotaxonomy and Mycotoxin production, *Mycologia* **81**, 837-861.

Frisvad, J.C., Filtenborg, O., Lund, F., and Thrane, U. (1992) New Selective Media for the Detection of Toxigenic Fungi in Cereal Products, Meat and Cheese, in R.A. Samson, Hocking, A.D., Pitt, J.I., and King, A.D. (eds.), *Modern Methods in Food Mycology*, Elsevier, Amsterdam, pp. 275-285.

Gams, W., van der Aa, H.A., van der Plaats-Niterink, A.J., Samson, R.A., and Stalpers, J.A. (1987) *CBS Course of Mycology*, CBS, Baarn.

Hult, K., Plestina, R., Habazin-Novak, V., Radic, B., and Ceovic, S. (1982) Ochratoxin A in Human Blood and Balkan Endemic Nephropathy, *Arch. Toxicol.* **51**, 313-321.

Kornerup, A. og Wanscher, J.H. (1969) *Farver i farver*, Politikens forlag, Kobenhavn.

Lund, F., Thrane, U., Frisvad, J.C., and Filtenborg, O. (1992) *IBT List of fungal cultures and their secondary metabolites*, 1st edition, Dept. of Biotechnology, The Technical University of Denmark.

Madhyastha, S., Marquardt, R.R., and Abramson, D. (1993) Effect of ochratoxin producing fungi on the chemical composition of wheat and barley, *Journal of Food Quality* 16, 287-299.

Pitt, J.I. (1979) *The Genus* Penicillium *and its Teleomorphic States* Eupenicillium *and* Talaromyces, Academic Press, New York.

THE PREVALENCE OF *FUSARIUM SOLANI* IN WRINKLED AND ROUND-SEEDED PEA GENOTYPES.

L. BØDKER
Danish Institute of Plant and Soil Science
Lottenborgvej 2, DK-2800 Lyngby
Denmark

Abstract

This study showed clearly that *F. solani* was present in nearly all lesions in wrinkled-seeded pea varieties for green harvest over a three year period. As expected, *F. solani* occurred in a much lower frequency in round-seeded varieties for dry peas. However, there was an interesting difference in aggressiveness of the different isolates according to the site of isolation on the plant which was reflected in the top-weight. This study shows the importance of pathogenicity test in connection to disease surveys, especially of root diseases. Among nearly all species of root pathogens, saprophytic variants are very prevalent. This can lead to misinterpretation of the importance of a certain pathogen if not representative isolates are tested for pathogenicity.

1. Introduction

Worldwide disease surveys have clearly shown that root diseases of peas are caused by a complex of pathogenic fungi. The lack of effective fungicides against the entire root disease complex and the increasing awareness of the environmental side effects of the pesticides have made genetic resistance to the dominating method for disease control. However, breeding programmes are time consuming and based on long term strategies on the basis of information of the prevalence and importance of the major pathogens. *Fusarium solani* (Mart.) Sacc. and *F. solani* f.sp. *pisi* (Jones) Snyd. & Hans. are recognized as belonging to the most important root rot pathogens in pea in most pea growing areas [1,2,5,9,10,14,17].

Fusarium root rot, caused by *F. solani* f.sp. *pisi*, is associated to dry, warm soil environment. The fungus may cause seedling rot as well as root rot of older plants. Peas are infected soon after germination and any factor that stresses the plant predispose it to

D. F. Jensen et al. (eds.), Monitoring Antagonistic Fungi Deliberately Released into the Environment, 57–61.

Fusarium [9]. The incidence of fusarium root rot in both pea and bean has in greenhouse studies been directly correlated to the quantity of carbohydrates exudated by the germinating pea plant [6,7,8,12] which is higher for wrinkled-seeded than for round seeded varieties [11,12,15,16].

The aim with this paper is to show the prevalence of *F. solani* in wrinkled seeded and round seeded pea genotypes under field conditions.

2. Materials

A disease survey was performed in 1581 fields of commercial crops of wrinkled- and round seeded peas for seed and green harvest, respectively in 1989-1991. In 1989, only wrinkled-seeded genotypes were surveyed in two separate areas, i.e., Funen and Lolland-Falster. In 1990 and 1991, the survey also included round-seeded peas, which originated from farmers' fields and from small plots in field trials for evaluation of new round-seeded varieties of peas uniformly scattered throughout Denmark. 252 plant samples with clear symptoms of root rot was selected for further investigation of the presence of *F. solani*.

2.1 ISOLATION

Out of twenty randomly collected plants from each field, five symptomatic plants were selected for isolation of *F. solani*. The stem was removed at the second node; the remaining root system was washed by thoroughly rubbing (15-30 sec) with hand soap and then placed under running tap water for at least one hour. The root system was surface disinfected in 1.5 % sodium hypochlorite for two minutes and dried between pieces of sterile filter paper. From the interface between healthy and diseased tissue on both epicotyls and roots, small transverse sections were cut out and transferred to plates with PDA (Difco) and SNA-media (14) with 25 ppm tetracyclines (Sigma) and 50 ppm chloramphenicol (Sigma) added. After incubation at 24 C for 5 days, cultures were transferred to fresh plates of the two isolation media for further 8-12 days for identification. Cultures of PDA were incubated in diffuse day light in the laboratory; those on SNA were incubated in growth room at 24±1 C in black light (Philips TL 20W/09 N) in a 12 h day length.

Isolates were stored on sterilized soil at 5°C.

2.2 PATHOGENICITY TEST

The inoculation test was based on the two round-seeded varieties, Bodil and Solara. The inoculation procedure was similar to Haglund (4). To see the effect of root injury, half of the plants were root pruned according to Haglund (4). Only isolates from wrinkled-seeded pea genotyped in 1990 were tested.

The disease severity was quantitatively estimated as the percentage of discoloured root system and scored into six classes (0-5), 0 = no discolouration, 1 = 1 -10 % discolouration, 2 = 11 - 30 %, 3 = 31 - 60 %, 4 = 61 -90 % and 5 = 91- 100 % and dead plants. A disease severity index was calculated separately for the epicotyl and the root of each plant. The two score totals for all twenty plants were added and divided by two to give a total disease severity index (0-100).

The top weight per plant was measure after drying the leaves thoroughly between several layers of tissue paper.

3. Results

There was no difference in susceptibility between the two round seeded varieties, Bodil and Solara, and no effect of root pruning.

F. solani was isolated from nearly all lesions of root rot in wrinkled-seeded pea genotypes and occurred in much lower frequence in round-seeded pea genotypes (Table1).

TABLE 1. The number of fields with wrinkled- and round-seeded pea varieties surveyed and the percentage from which *Fusarium solani* as been isolated from symptomatic plants.

	1989		1990		1991	
	Number	Per cent	Number	Per cent	Number	Per cent
Wrinkled-seeded	95	94	30	78	70	80
Round-seeded	-[a]	-[a]	12	25	45	24

[a] No fields of round-seeded varieties were surveyed in 1989

The pathogenicity test showed a close correlation between disease severity index and top weight. The strains isolated from hypocotyl and the adventitious roots were often less pathogenic than the strains isolated from epicotyl (Figure 1).

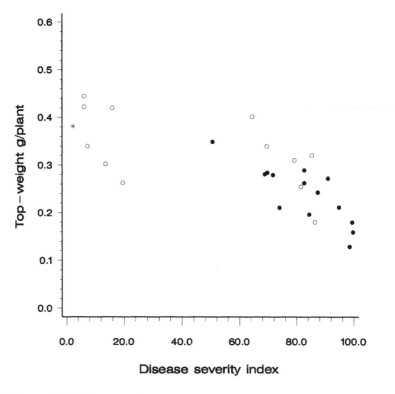

Figure 1. The correlation between disease severity index (0-100) and the yield (g/plant) of the canopy in greenhouse trials when different isolates of *Fusarium solani* were tested for pathogenicity. Site of isolation: ✳✳✳ control, ●●● epicotyl, ○○○ root.

4. Discussion

This study showed clearly that *F. solani* was present in nearly all lesions in wrinkled-seeded pea varieties for green harvest over a three year period. As expected, *F. solani* occurred in a much lower frequency in round seeded varieties for dry peas. This was in agreement with earlier reports which have shown that the activity of damping-off and root rot causing organisms is stimulated by the higher exudation of carbo-hydrate in wrinkled than in smooth-seeded peas [3,7,8,11,12,16]. However, there was an interesting difference in aggressiveness of the different isolates according to the site of isolation on the plant which was reflected in the top weight. In general the strains isolated from the epicotyl were more aggressive than the strains isolated from the roots (Figure 1). The surface disinfection seems to be more efficient on the smooth epicotyl than on the rugged hypocotyl. Despite the thoroughly rubbing of the roots with hand soap and the disinfection with sodium hypochlorite, it seems not possible to avoid saprophytic strains on the roots.

This study shows the importance of pathogenicity test in connection to disease surveys, especially of root diseases. Among nearly all species of root pathogens, saprophytic variants are very prevalent which can lead to misinterpretation of the importance of a certain pathogen if not representative isolates are tested for pathogenicity. The distribution of saprophytic and pathogenic strains of *F. solani* seems not only to be different between fields but even different within the root system.

5. References

1. Basu, P.K., Crête, R, Donaldson, A:G.,Gourley, C.O., Haas, J.H., Harper, F.R., Lawrence, C.H., Seaman, W.L., Toms, H.N.W., Wong, S.I. and Zimmer, R.C.: Prevalence and severity of diseases of processing peas in Canada, 1970-71. *Can. Plant Dis. Surv.* **53** (1973), 49-57.
2. Biddle, A.J.: A prediction test for pea footrot and the effects of previous legumes. *Proc. Brit. Crop. Protect. Conf. Pest. Dis.* (1984), 773-777.
3. Flentje, N.T. and Saksena, H.K.: 1964. Pre-emergence rotting of peas in south Australia. *Aust. J. Biol. Sci.* **17** (1964), 665-75.
4. Haglund, W.A.: A rapid method for inoculating pea seedlings with *Fusarium oxysporum* f.sp. *pisi*. *Plant Dis.* **73** (1989), 457-458.
5. Hwang, S.F. and Chang, K.F.: Incidence and severity of root rot disease complex of field pea in northeastern Alberta in 1988. *Can. Plant Dis. Surv.* **69** (1989), 139-141.
6. Kraft, J.M.: The influence of seedling exudates on the resistance of peas to Fusarium and Pythium root rot. *Phytopathology* **64** (1973), 190-193.
7. Kraft, J.M.: The role of delphinidin and sugars in the resistance of pea seedlings to fusarium root rot. *Phytopathology* **67** (1977), 1057-1061.
8. Kraft, J.M.: Seed electrolyte loss and resistance to fusarium root rot of peas. *Plant Dis.* **70** (1986), 743-745.
9. Kraft, J.M., Burke, D.W., and Haglund, W.A.: Fusarium diseases of beans, peas, and lentils. Page 142-156 in: *Fusarium: Diseases, Biology, and Taxonomy.* P.E. Nelson, T.A. Toussoun, R.J. Cook, eds. The Pennsylvania State University Press, University Park & London, 1981.
10. Mabey, M. and Whalley, WM: Experimental analysis of the foot rot complex of peas. *Brighton Crop Protection Conference-Pest and Diseases* (1988), 629-636.
11. Matthews, S. and Whitbread, R: Factors influencing pre-emergence mortality in peas. 1. An association between seed exudates and the incidence of pre-emergence mortality in wrinkle-seeded peas. *Plant Pathol.* **17** (1968), 11-17.
12. Muehlbauer, F.J. and Kraft, J.M.: Effect of pea seed genotype on pre-emergence damping-off and resistance to *Fusarium* and *Pythium* root rot. *Crop Sci.* **18** (1978), 321-323.
13. Nirenberg, H.I.: Untersuchungen über die morphologische und biologische differenzierung in der *Fusarium*-Sectionen *Liseola*. *Mitt. Biol. Bundesanst. Land Forstwirtsch.* **127** (1976), 64-67.
14. Oyarzun, P., Gerlagh, M. and Hoogland, A.E.: Pathogenic Fungi Involved in Root Rot of Peas in the Netherlands and Their Physiological Specialization. *Neth. J. Pl. Path.* **99** (1993), 23-33.
15. Schroth, M.N. and Cook, R.J.: Seed exudation and its influence on pre-emergence damping-off of bean. *Phytopathology* **54** (1976), 670-673.
16. Short, G.E. and Lazy, M.L.:. Factors affecting pea seed and seedling rot in soil. Phytopathology 66, 188-192.
17 Tu, J.C. Incidence and etiology of pea rots in southwestern Ontario. *Can. Plant Dis. Surv.* **66** (1986), 35-36.

RELEASE OF A CORD-FORMING BASIDIOMYCETE ANTAGONISTIC TO HETEROBASIDION ANNOSUM AND ITS SUBSEQUENT REISOLATION AND IDENTIFICATION

L. HOLMER & J. STENLID
Department of Forest Mycology and Pathology
Swedish University of Agricultural Sciences
Box 7026, S-750 07 Uppsala, Sweden

Abstract

After initial competition experiments done in the laboratory, the most successful species, Resinicium bicolor (Alb. & Schw. ex Fr.) Parm., was chosen as a potential control agent of Heterobasidion annosum (Fr.)Bref (Holmer & Stenlid, 1993). Three different field experiments in first rotation field plantations of Norway spruce (Picea abies) were conducted in three places in Sweden. Wood blocks, preinoculated with one strain of R. bicolor, were buried in the soil beside stumps at different time interval after thinning and with different number of wood blocks. In two of the experiments, half of the stumps were treated with a suspension of H. annosum conidia from one strain sprayed on the surface. After 2-3 years, stump roots were investigated and the length of growth of both species were noted. Reisolations of mycelia were made from the roots as well as from wood debris in the experiment area. The isolates were paired on agar plates against the original strains (somatic incompatibility test). The used strain of R. bicolor was found again all over the experimental areas, no other strains of the species were found. The used strain of H. annosum was only reisolated from roots of artificially H. annosum treated stumps. Other strains of the fungus were frequently found from the untreated stumproots. R. bicolor was found to have limited effect on the growth and occurence of H. annosum.

1. Introduction

Root rot caused by Heterobasidion annosum is, in proportion to the area, a greater problem in first-rotation field plantations of Norway spruce (Picea abies) than on old forest sites. Damages on the trees usually appears after the first thinning operation and may in field plantations result in death of trees in addition to root and butt rot (Swedjemark & Stenlid, 1993). One reason for being so could be that competing saprophytic fungi are lacking in field plantations (Kirby et al., 1990). These fungi could possibly, through invading dead or dying spruce roots, prevent the spreading of root rot via root contacts between stumps and

63

D. F. Jensen et al. (eds.), Monitoring Antagonistic Fungi Deliberately Released into the Environment, 63–70.
© 1996 Kluwer Academic Publishers. Printed in the Netherlands.

healthy trees. If so, it may be possible to reduce the damages in field plantations by introducing suitable competing fungi as biological control.

When introducing a fungus as a biological control agent it is important to be able to reisolate and identify the used strain after some time. Many species of Basidiomycotina have been shown to form reaction zones when two different genets; i.e. genetically distinct secondary mycelia; (Rayner and Boddy, 1988) are inoculated in the same Petri dish. Examples are Phellinus tremulae (Holmer et al. 1994), H. annosum (Stenlid, 1985) and Hymenochaete tabacina (Stenlid and Holmer, 1991). The method is easy and relatively accurate, and is therefore of great use in studies like this.

In this study we introduced a cord-forming basidiomycete, Resinicium bicolor, in first rotation plantations of Norway spruce, Picea abies. This species was chosed after initial laboratory studies, made as competition tests, where it had considerable success (Holmer and Stenlid, 1994; Holmer and Stenlid, unpublished). The aim was to verify if R. bicolor could be efficient in checking H. annosum in the forest, and also to find a suitable method for introducing the fungus in the soil of forest plantations.

2. Materials and methods

2.1. STRAINS AND CULTURE MEDIA

The two used species, R. bicolor and H. annosum were both represented by a single strain and isolated from stumps of Norway spruce in a 100-year old conifer forest near Uppsala. The strains were cultured on Petri dishes with Hagem agar (HA) (Stenlid, 1985) at 20°C until used in the experiments. Conidia suspension of H. annosum were prepared from three weeks old plates with numerous amount of conidia. Each plate were rinsed with 40 ml of sterile water, immediately before use.

2.2. PREPARATION OF WOOD INOCULUM

15 mm thick branches of Norway spruce were cut into 40 mm long pieces, which were autoclaved twice at 120 °C at a 24 hour interval. The wood pieces were placed on HA agar in 250 ml E-flasks and inoculated with R. bicolor for one month at room temperature.

2.3. EXPERIMENTAL PROCEDURE

Field plantations of Norway spruce at three locations in Sweden; Bälinge, Asa and Remningstorp, were chosen for the field studies. The plantations were all 30 years of age and thinned for the first time when the experiments were carried out.

2.3.1. Bälinge
In a randomized block design, 25-38 stumps were infected monthly for four months by burying one wood block inoculated with R. bicolor in the soil beside a marked root 30 cm from each spruce stump used in the study. The first inoculation was made one month after the thinning.

2.3.2. Asa

One wood block was buried in the soil beside one root 20 cm from the stump immediately after thinning. The outer layer of the bark was removed from the root at the site of contact. The experiment contained 32 blocks, each with two stumps. Two roots per stumps were marked; one was infected with R. bicolor. On the surface of one stump per block a conidia suspension of H. annosum was sprayed.

2.3.3. Remningstorp

Two stands were included in the experiment; one planted on old agriculted field and one old forest site. A ring of wood blocks inoculated with R. bicolor were buried in the soil around the stump at two different distances, 30 and 70 cm. Eight wood blocks were used at the 30 cm distance and sixteen at the 70 cm distance. The inoculations were made immediately after the thinning. Fifteen blocks were used, each containing six stumps with different treatment. The stump surface of threee of these stumps were sprayed with a conidia suspension of H. annosum. The same design was used for the two stands.

2.3.4. Harvesting

In Bälinge, the marked root were collected three years after the inoculations. The two marked roots on each stump and one root per stump were collected after two years in Asa and Remningstorp, respectively.

The collected roots were carefully removed, marked and brought to the laboratory in separate plastic bags. In the laboratory, the roots were cut into 10 cm long pieces, which were incubated in plastic trays for five days. The root pieces were kept moist under the incubation time. Conidia of H. annosum and mycelia of R. bicolor were identified at 40 × magnification by using a dissecting microscope.

2.3.5. Somatic incompatibility tests

Samples from cords of R. bicolor found on brushwood and branches in the area were taken and brought to the laboratory for isolation. Samples from cords of R. bicolor and from conidia of H. annosum were also taken from the investigated roots. The isolations were made on HA plates with 5 ppm Benlate.

The obtained mycelia from both species were thereafter paired against the original strains on 1.5 % MA plates. The plates were incubated in darkness at room temperature for two weeks where after checking for reaction zones was made.

3. Results

3.1. SOMATIC INCOMPATIBILITY TESTS

When investigating the surronding brushwood and branches in the study areas, it was found that R. bicolor had had a great success, except in Bälinge. In Asa and Remningstorp, cords of the fungi could be found all over the area, on and under piles of brush woods and under branches and logs. Mycelia could also be found on and under the bark of the stumps and, in

Asa, there were furthermore quite a large amount of fruibodies. The samples taken from cords from brushwoods was of the same genotype as the one used in the experiment.

The used strain of H. annosum could only be refound from the conidia sprayed stumps. "Wild" strains was isolated both from the conidia treated stumps and from those without treatment. Many roots contained both H. annosum and R. bicolor (Fig. 1).

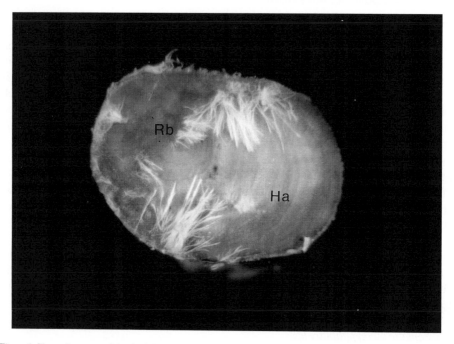

Figure 1. Piece of root containing both R. bicolor and H. annosum.

3.2. INVESTIGATED ROOTS

3.2.1. *Bälinge*

Few stumps were naturally infected by H. annosum at this site; on average for all four inoculation times only 16 % of the investigated roots were found infected. The average length of growth in the roots was less on the earliest inoculated roots; on the other hand the number of roots infected were higher in those roots. An average of 72 % of the roots contained R. bicolor and the average length of growth was greater in the earliest inoculated roots. The number of infected roots were also higher after one month than in later inoculations.

The average length of growth of H. annosum in roots with R. bicolor was less in the early inoculated roots than in the later ones. It was also less compared with the roots where only H. annosum was found (Table 1).

TABLE 1. The total number of roots, number of roots with H. annosum and R. bicolor alone and combined with eachother, the average length of growth of H. annosum and R. bicolor alone and combined with eachother in Bälinge.

Time after thinn., months	Tot. no of roots	Average length of growth of Ha, cm	Average length of growth of Ha-Rb, cm	Average length of growth of Ha+Rb, cm	Average length of growth of Rb, cm	Average length of growth of Rb-Ha, cm	Average length of growth of Rb+Ha, cm	No of roots with Ha, %	No of roots with Rb, %
1	25	32	50	20	55	54	60	20	88
2	38	48	60	46	42	42	38	16	76
3	34	52	40	55	36	30	38	15	65
4	38	46	-	46	35	36	32	13	60

3.2.2. Asa

The percentage of H. annosum infections were less on the roots were R. bicolor had been inoculated. This was true whether the stumps were naturally or artificially infected with H. annosum (Fig. 2a). However, when the infection of H. annosum had reached the roots, the average length of growth was greater on the R. bicolor inoculated roots, at least on the naturally infected stumps (Fig. 2b).

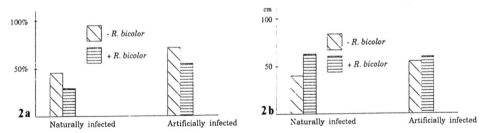

Figure 2. a) Percentage of H annosum infected roots in Asa, with and without R. bicolor. b) Average length of growth of H. annosum with and without R. bicolor.

3.2.3. *Remningstorp*

More roots were infected with R. bicolor than had been inoculated, especially so in the field plantation. There were also many natural infections of H. annosum, most of them in the forest plantation (Fig. 3a).

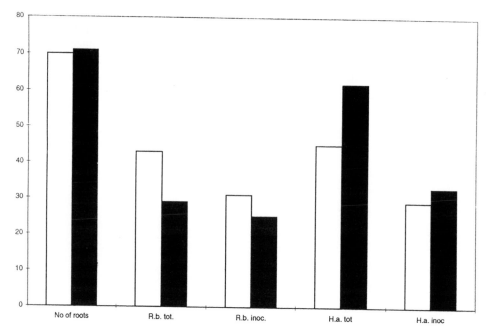

Figure 3a). The total number of roots, total number of roots infected with H. annosum and R. bicolor and number of roots inoculated with H.annosum and R. bicolor in field- and forest plantations in Remningstorp. White staples-field plantation. Black staples-forest plantation.

The average length of growth for H. annosum was higher in the forest plantation; for R. bicolor it was highest in the field plantation. The average length of growth of H. annosum without R. bicolor was somewhat higher (61.7 and 64.2 cm field and forest respectively) than with R. bicolor (53.7 and 61.2 cm). The average length of growth of R. bicolor without H. annosum in the forest plantation was much less (30 cm) than with H. annosum (57.7 cm). The situation was the same in the field plantation, but to a lower degree (Fig.3b).

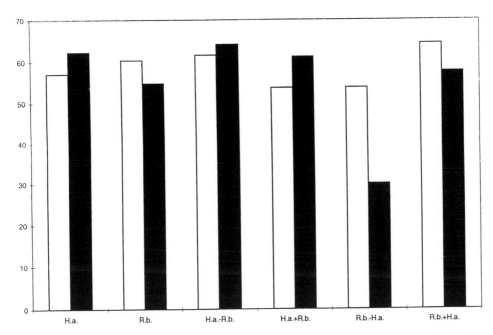

Figure 3b) Average length of growth in cm of H. annosum *and* R. bicolor *alone and combined with eachother in field-and forest plantations in Remningstorp. White staples-field plantation. Black staples-forest plantation.*

4. Discussion

The biological (and chemical) control of H. annosum has up to now mostly been concentrated to prevent spore germination on the surface of fresh stumps. The reducing effect of Peniophora gigantea on the infections of H. annosum on pine has been well documented (Risbeth, 1963; Greig, 1975), and has resulted in widespread use in pine forests in Britain. Lately, P. gigantea has also been used to prevent the spread of H. annosum of stumps of Norway spruce (Korhonen, pers. communication).

The idea of preventing the spread of H. annosum via root contacts is a new approach of biological control. One problem with this idea is to find an easy way to introduce the antagonistic fungus into the soil. The fungus should also be able to spread itself from the inoculum. Studies by Boddy (1993), showed that cord-forming basidiomycetes are able to effectively spread through non-sterile soil, and this fact make this type of fungi highly interesting for purposes like this. Kirby et al. (1990), studied R. bicolor and found it efficient in colonizing thinning stumps and observations indicated that it might be a competitor of H. annosum in dying rootwood in forest soil. In laboratory experiments, using cord forming species in interaction tests against H. annosum, R. bicolor was found to be the most efficiant in replacing the pathogen (Holmer and Stenlid, 1994; Holmer and Stenlid, unpublished).

The results from this study confirms the observations made by Kirby et al (1990). R. bicolor was, at least in two of the experimental areas, highly efficiant in spreading itself from the inoculated wood blocks. It was also able to invade and spread itself through the spruce roots, in some cases more than 90 cm in two years.

The results from both Asa and Remningstorp shows that R. bicolor had some success in depressing the growth of H. annosum; in Remningstorp it was most efficiant in the field plantation. One reason for the differentiation between field- and forest plantation could be that in field plantations other competing decomposing fungi are rare, and thus R. bicolor did not have to defend its substrate.

The results from Bälinge showed that R. bicolor was more efficiant in root growth if it was inoculated soon after thinning. This could indicate that competing fungi may have invaded the area. Some other wood decomposing species, like Coniophora sp. was also isolated from the collected roots.

Unfortunately, R. bicolor did not have the wanted efficiency in preventing the spread of H. annosum. One possible explanation could be that the stump roots are still alive some time after the felling of the tree, and since R. bicolor only is weakly pathogenic (Holmer, unpublished), it might take some time for it to invade the roots. This time delay may be enough to allow H. annosum to grow through the roots. A cord-forming species that is slightly more pathogenic could perhaps be more efficient.

5. References

Boddy, L. (1993) Saprotrophic cord-forming fungi: Warfare strategies and other ecological aspects. *Mycol. Res.* 97, 641-655.

Greig, J.W. (1975) Chemical, biological and silvicultural control of Fomes annosus. In: Proceedings of the Fifth International Conference on Problems of Root and Butt Rot in Conifers.

Holmer, L., and Stenlid, J. (1993) The importance of inoculum size for the competitive ability of wood decomposing fungi. *FEMS Microb. Ecol.* 12, 169-176.

Holmer, L., Nitare, L., and Stenlid, J. (1994) Population structure and decay pattern of Phellinus tremula as determined by somatic incompatibility. *Can. J. Bot.* 72, 1391- 1396.

Holmer, L., and Stenlid, J. Diffuse competition for heterogenous substrate in soil among wood decomposing basidiomycetes. Unpublished.

Kirby, J.J.H., Stenlid, J., and Holdenrieder, O. (1990) Population structure and responses to disturbance of the basidiomycete Resinicium bicolor. *Oecologia* 85, 178-184.

Rayner, A.D.M., and Boddy, L. (1988) Fungal decomposition of wood. John Wiley & Sons, Chichester, England.

Risbeth, J. (1963) Stump protection against Fomes annosus. III. Inoculation with Peniophora gigantea. *Ann. appl. Biol.* 52, 63-77.

Stenlid, J. (1985) Population structure of Heterobasidion annosum as determined by somatic incompatibility, sexual incompatibility and isoenzyme patterns. *Can. J. Bot.* 63, 2268-2273.

Stenlid, J., and Holmer, L. (1991) Infection strategy of Hymenochaete tabacina. *Eur. J. For. Path.* 21, 313-318.

Swedjemark, G., and Stenlid, J. (1993) Population dynamics of the root rot fungus Heterobasidium annosum Following thinning of Picea abies. *Oikos* 66, 247-254.

METHODS TO QUANTIFY NEMATOPHAGUS FUNGI IN SOIL: MICROSCOPY OR GUS GENE ACTIVITY.

LOTTA PERSMARK, YVONNE PERSSON
AND HANS-BÖRJE JANSSON
Dept of Microbial Ecology, Lund University, Ecology Building,
S-223 62 Lund, Sweden.

1. Introduction

Nematodes are common soil inhabitants. Some feed on microorganisms and others are parasites of plants and animals (Stirling, 1991). Nematophagous fungi are parasites of nematodes and are found in almost every soil. They can be divided into three groups: 1) The nematode-trapping fungi, which capture nematodes by means of hyphal traps. These fungi live to varying degrees as saprophytes. 2) The endoparasitic fungi, which parasitize nematodes using minute conidia or zoospores and are obligate parasites (Barron, 1977; Gray, 1988). 3) Egg and female parasites which use their hyphal tips to infect the nematodes. These fungi live also as saprophytes (Morgan-Jones and Rodriguez Kabana, 1988).

Little is known about the ecology of these fungi. Most studies are restricted to surveys on the occurrence of nematophagous fungi in different habitats and in different climates (Gray, 1988). Only few quantitative studies have been performed (Dackman *et al.*, 1987). Jaffee *et al.* (1992) showed that a density dependent relationship exists between the endoparasitic fungus *Hirsutella rhossiliensis* and its nematode prey. The nematophagous fungi that live partly as saprophytes might have a more complex relation to their nematode host, and their ability to regulate nematode numbers in soil is not known.

Plant-parasitic nematodes are important pests of crops and since many nematicides have been banned due to environmental hazards, the search for alternative control methods is intensive. The possibility of using nematophagous fungi in biological control of plant-parasitic nematodes have increased the interest in these fungi. To be successful in such attempts much more research have to be performed on the ecology of these fungi in natural soils (Stirling, 1991).

In most studies using nematophagous fungi as biocontrol agents towards plant-parasitic nematodes no determination of the establishment of the added fungus have been performed. Mainly indirect methods, such as the reduction in nematode numbers, have been used to determine the activity of the added fungi. A broader knowledge of the establishment of fungi added to soil will be important in the search for successful biocontrol agents.

Arthrobotrys oligospora is one of the most commonly occurring nematode-trapping fungi. It is common especially in agricultural soils. (Duddington, 1954; Shepherd, 1956). This fungus captures nematodes by means of adhesive three-dimensional network

D. F. Jensen et al. (eds.), Monitoring Antagonistic Fungi Deliberately Released into the Environment, 71–75.
© *1996 Kluwer Academic Publishers. Printed in the Netherlands.*

traps formed on the hyphae, but conidia exposed to certain environmental conditions form traps also immediately upon germination (Dackman and Nordbring-Hertz 1992). Apart from its nematode-trapping ability, it is regarded as a fairly good saprophyte. Furthermore, *A. oligospora* can also feed on other fungi, and it is known to be a mycoparasite of *Rhizoctonia solani* and other fungi (Persson *et al.*, 1985)

Detection and quantification of nematophagous fungi in soil are quite laborious tasks. Therefore, new techniques have to be developed for studies of these (and other) fungi in soil. The needs for specific markers are high. Methods to transform fungi are now available (Roberts *et al.*, 1989). ß-Glucuronidase (GUS) is a commonly used reporter enzyme for plant and fungal genetic research. Advantages of using GUS as a reporter gene is among others that most plant and fungal systems lack appreciable GUS-activity, and that GUS assay substrates are available for histochemical, spectrophotometric and spectrofluorometric analyses (Galagher, 1992)). In soil, GUS activity probably is low. In fresh waters GUS activity measurements is used to detect coliform bacteria (Rice, 1991).

In order to find a reliable method to quantify nematophagous fungi in soil, we compared two methods: traditional soil dilution with a most probable number estimation and spectrofluorometrical detection of a GUS transformed *A. oligospora*.

2. Material and methods

2.1. MPN ESTIMATIONS

A soil dilution method with a most probable number (MPN) estimation according to Dackman *et al.* (1987) was performed. Soil was mixed with 0.01% Calgon (sodium hexametaphosphate) and a dilution series was made. From each dilution 1 ml was spread on 5 replicate water agar plates baited with 1000 nematodes. The plates were screened after 1 week under a compound microscope for the presence of nematophagous fungi. A MPN-table with 5-fold dilutions with 5 plates per dilution was constructed after Halvorson and Ziegler (1933). In this way we have followed the population dynamics of nematode-trapping fungi in agricultural soils and quantified the density of nematode-trapping fungi in the rhizosphere of agricultural crops grown in field or in pots. Furthermore, we have followed the establishment of nematophagous fungi added to soil in microcosm experiments using either unsterilized soil or soil heated in a microwave oven until the temperature reached 100°C.

2.2. GUS TRANSFORMATION

A GUS-producing strain of *A. oligospora* was created by cotransformation of protoplasts by two different plasmids (Persson *et al.*, 1995): pAN-7.1, carrying the *Escherichia coli* hygromycin B phosphotransferase gene as a dominant selectable marker, and the gpd promoter and the trpC terminator from *Aspergillus nidulans* (Punt *et al.*, 1987) and pNOM 102, carrying the gene for GUS expression (Roberts *et al.*, 1989). Protoplasts of *A. oligospora* were produced by treatment of mycelia with Novozym 234 (10 mg/ml) and Lyticase (1 mg/ml). Transformation was carried out by

PEG/CaCl$_2$ mediated uptake of the plasmids into the protoplasts. Protoplasts were allowed to regenerate in STC (1.2 M Sorbitol, 10 mM TRIS -HCl, pH 7.5, 50 mM CaCl$_2$, 35 mM NaCl) for three days before cultivation on malt extract agar containing hygromycin (100 µg/ml). Successful cotransformation was confirmed by addition of 2 mM X-gluc (5-bromo-4-chloro-3-indyl glucuronide) directly onto the colonies, where the GUS-positive colonies were stained blue. Single spore isolations were repeated five times.

Transformants were compared with the wild-type strain in radial growth rate on corn meal agar (CMA, Difco) diluted 10 times, growth in liquid culture in 0.5% soya peptone (neutralized, Oxoid) at 20°C on a rotary shaker, spore production at 20°C on CMA supplemented with 2 g/l K$_2$HPO$_4$ and trap formation on SP+phe-val (0.04% soya peptone (neutralized, Oxoid), 0.01% phenylalanyl-valine (phe-val, Sigma) and 1% agar) (Nordbring-Hertz et al., 1995).

2.3. GUS ACTIVITY MEASUREMENT

The transformed A. oligospora was grown in 0.5% soya peptone (neutralized, Oxoid) at 20°C on a rotary shaker. GUS activity in pure culture was measured spectrofluorometrically using MUG (4-methylumbelliferyl ß-D-glucuronide) as substrate according to Gallagher (1992). Protein was measured according to Bradford (1976). To measure GUS-activity in soil an extraction was first performed: ten g soil and 10 ml extraction buffer was mixed in a sorvall omnimixer at 16 000 rpm for 1 min. After a few min sedimentation the extract was centrifuged at 1 000 g for 10 min and the supernatant was used for GUS-activity measurements as above.

Detection limits for spores and hyphae added to a non-sterilized sandy agricultural soil was determined. GUS-activity was measured immediately after addition of the fungus to the soil.

2.4. QUANTIFICATION OF THE TRANSFORMANT

Spores of the transformed A. oligospora was pelletized in alginate (10 000 spores/pellet) according to Lackey et al. (1993) One pellet/g soil was added to an agricultural soil, which contained a natural flora of nematophagous fungi, and to a peat-sand mix containing no such fungi. The soils were incubated at 15°C and analysed for the presence of the A. oligospora transformant after 1, 3, 7 and 14 days using both the MPN and GUS assays.

3. Results and discussion

It was possible to follow population dynamics of nematode-trapping fungi both in soil, rhizosphere and in microcosm systems using the MPN-method. The densities of nematode-trapping fungi in agricultural soils varied between 1 and 200 propagules per g soil depending on the soil depth and the time of the year. The highest densities were found in the upper 20 cm in the end of the growing season. In the rhizosphere up to 5 000 propagules per g root were recorded. In microcosms using partially sterilized soil up to 300 000 propagules per g soil were recorded after addition of both nematodes, green manure and pelletized *A. oligospora*.

The growth rate (radial and in liquid culture), spore production and trap formation of the transformants was similar to the wild type. The GUS-activity was 5 nmol/mg·min in ungerminated spores and ca 25 nmol/mg·min in mycelia from 5 day old cultures grown in 0.5% soya-peptone on a rotary shaker. This was 4 times less than figures reported for the GUS-transformed fungus *Bipolaris sorokiniana* (Liljeroth *et al.*, 1993).

Detection limits for spore and hyphal suspensions were approx. 100 000 spores or 3 mg hyphae per g soil respectively. These amounts of fungi are not expected to be found under natural conditions in soil. The background GUS-activity were probably high as the agricultural soil was fertilized using a compost containing manure and offal. In other soils it might be possible to detection smaller amounts of fungi.

It was not possible to detect any GUS-activity above background in either of the two soils after addition of pellets of the transformed *A. oligospora*. The establishment of the fungus was, however, detected 3 days after addition of the pellets using the MPN-method. After 14 days the density in the peat-sand mix was 900 propagules per g. We do not yet know how the density of fungi recorded with the MPN method corresponds to the GUS-activity measured spectrofluorometrically.

4. Conclusions

The MPN method was successful in showing a variation in the occurrence of nematophagous fungi in natural systems, and also to distinguish between the different species present. Verification of the growth of *A. oligospora* added to non-sterilised soil was possible with the MPN method. The GUS-method have so far not proven to be as efficient in quantifying the GUS-transformed *A. oligospora* in non-sterilised soil as the MPN-method. The technique to measure the GUS-activity directly in soil has to be further improved and work in this direction is carried out in our laboratory. The usefulness of the transformed *A. oligospora* for ecological studies in soils is still unclear. The transformant will probably be an important tool for the study of nematophagous fungi in the rhizosphere where the densities of nematophagous fungi are higher than in soil. There is also a possibility of using soils with less background activity.

4. References

Barron, G.L. (1977) *The nematode-destroying fungi. Topics in mycobiology No.1.* Canadian Biological Publications Ltd., Guelph, Ontario Canada.

Bradford, M.M. (1976) A rapid and sensitive method for the quantitation of microgram quantities of protein utilizing the principle of protein-dye binding. *Anal. Biochem.* **72**, 248-254.

Dackman, C. and Nordbring-Hertz, B. (1992) Conidial traps - a new survival structure of the nematophagous fungus *Arthrobotrys oligospora. Mycol.res.* **96**, 194-198.

Dackman, C., Olsson, S., Jansson, H.-B., Lundgren, B. and Nordbring-Hertz B. (1987) Quantification of predatory and endoparasitic nematophagous fungi in soil. *Microb. Ecol.* **13**, 89-93.

Duddington C.L. (1954) Nematode-destroying fungi in agricultural soils. *Nature* **173**, 500-501.

Gallagher, S.R. (1992) Quantitation of GUS Activity by Fluorometry pp. 47-59 in *GUS Protocols* (R.S. Gallagher Ed.) Academic Press, Inc. San Diego.

Gray, N.F. (1988) Fungi attacking vermiform nematodes. pp. 3-38 in *Diseases of nematodes*, vol. 2. (G.O. Poinar Jr, and H.-B. Jansson, Eds.) CRC Press, Boca Raton.

Halvorson, H.O. and Ziegler, N.R. (1933) Application of statistics to problems of bacteriology. II. A consideration of the accuracy of dilution data obtained by using a single dilution. *J. Bacteriol.* **26**, 520-530.affee, B.A., Phillips, R., Muldoon, A. and Mangel, M. (1992) Density-dependent host-pathogen dynamics in soil microcosms. *Ecology* **73**, 495-506.

Jeffersson, R.A. (1987). Assaying chimeric genes in plants: The GUS gene fusion system. *Plant Mol. Bio. Rep.* **5**:387-405.

Lackey, B.A., Muldoon, A.E. and Jaffee, B.A. (1993) Alginate pellet formulation of *Hirsutella rhossiliensis* for biological control of plant-parasitic nematodes. *Biological Control* **3**, 155-160.

Liljeroth, E., Jansson, H.-B. and Schäfer, W. (1993) Transformation of *Bipolaris sorokiniana* with the GUS gene and use for studying fungal colonisation of barley roots. *Phytopathology* **83**, 1484-1489.

Morgan-Jones, G. and Rodriguez Kabana, R. (1988) Fungi colonizing cysts and eggs. pp. 39-58 in *Diseases of nematodes*, vol. 2. (G.O. Poinar Jr, and H.-B. Jansson, Eds.) CRC Press, Boca Raton.

Nordbring-Hertz, B., Neumeister, H., Sjollema, K. and Veenhuis, M. (1995) A conidial trap-forming mutant of *Arthrobotrys oligospora*. *Mycol. Res.* in press).

Persmark, L. and Jansson, H.-B. (1995) Nematophagous fungi in the rhizosphere of agricultural crops. *Nematropica* (in preparation).

Persson, Y., Veenhuis, M. and Nordbring-Hertz, B. (1985) Morphogenesis and significance of hyphal coiling by nematode-trapping fungi in mycoparasitic relationships. *FEMS Microbiol. Ecol.* **31**, 283-291.

Persson, Y., Persmark, L., van den Hondel, C.A.M.J.J., Punt, P. and Jansson. H.-B.(1995) Transformation of the nematode-trapping and mycoparasitic fungus *Arthrobotrys oligospora* (in preparation).

Punt, P.J., Oliver, R.P., Dingemanse, M.A., Pouwels, P.H. & van den Hondel, C.A.M.J.J. (1987) Transformation of Aspergillus based on the hygromycin B resistance marker from *Escherichia coli. Gene* **56**, 117-124.

Rice, E.W., Allen, M.J., Brenner, D.J. and Edberg, S.C. (1991) Assay for ß-Glucuronidase in species of the genus *Escherichia coli* and its application for drinking water analysis. *Appl. Environ. Microbiol.* **57**, 592-593.

Roberts, I.N., Oliver, R.P., Punt, P.S. and van den Hondel, C.A.M.J.J. (1989) Expression of the *E. coli* ß-glucuronidase gene in industrial and phytopathogenic filamentous fungi. *Current Genetics* **15**, 177-180.

Shepherd, A.M. (1956) A short survey of Danish nematophagous fungi. *Friesia* **5**, 396-408.

Stirling, G.R. (1991) *Biological control of Plant Parasitic Nematodes*. CAB International, Wallingford.

MONITORING OF A GUS TRANSFORMED STRAIN OF *TRICHODERMA HARZIANUM* IN SOIL AND RHIZOSPHERE

HELGE GREEN, AND DAN FUNCK JENSEN
Department of Plant Biology, Plant Pathology Section, The Royal Veterinary and Agricultural University, Thorvaldsensvej 40, DK-1871 Frederiksberg C, Copenhagen, Denmark.

1. Abstract

Recently molecular technology has enabled reporter and marker genes to be integrated in the fungal genome. The resulting transformants can then be monitored after introduction into natural environments. This paper summarizes the suitability of using a GUS transformant for monitoring the presence, the population development, and the activity of a specific strain of *T. harzianum* deliberately released into the environment.

2. Introduction

The fungal antagonist *Trichoderma harzianum* Rifai, has been shown to be an effective biocontrol agent against a range of important aerial and soil-borne plant pathogens [4, 12, 15]. In order to obtain a reliable effect of the organism under different environmental conditions we need to know more about the ecology of the antagonist.

In microbial ecology numerous methods for assessing populations, activity, and biomass have been developed [18], but a recurrent problem has been the inability to detect, monitor, and recover specific microorganisms either naturally present or deliberately released into the environment. This has become even more of a concern now that the regulatory agencies of many countries require that microbial biocontrol products, containing either naturally occurring organisms or those that are genetically modified, have to be assessed in terms of their risk to the environment into which they are to be released.

With respect to *Trichoderma* spp., dilution plating on selective media has often been used for quantitative isolation from soil [6, 7, 17]. However, a major drawback to this method is that the majority of colonies arise from conidia rather than hyphae [23] which makes a correlation to fungal activity difficult. Even on selective media problems with the background microflora are common, and in all cases it is difficult to distinguish between introduced and indigenous *Trichoderma* spp. These problems have been partially overcome by utilizing fungicide resistant strains [1, 2, 16].

D. F. Jensen et al. (eds.), Monitoring Antagonistic Fungi Deliberately Released into the Environment, 77–83.
© 1996 *Kluwer Academic Publishers. Printed in the Netherlands.*

More recently, molecular approaches have enabled reporter and marker genes to be integrated in the fungal genome. Such strains can then be monitored after introduction into natural environments. The GUS (ß-glucuronidase) gene from *E. coli* seems very promising as background activity is either absent or low in most plants and fungi investigated [11, 24]. Furthermore, the enzyme is fairly stable and can easily be assayed by different methods [9, 11]. GUS transformants of several plant pathogenic fungi have been used for detection and biomass quantification in infected plant tissue [5, 13, 14].

In this paper we discuss the use of the GUS gene as a marker for monitoring an introduced *T. harzianum* strain in the rhizosphere. The discussion are related to general methodologies for assessing microbial populations and biomasses. Experimental methods and detailed results are described elsewhere [10, 21].

3. Transformation of *T. harzianum*

Thrane et al. [21] cotransformed *T. harzianum* with the ß-glucuronidase gene (GUS) and the hygromycin resistance gene (Hyg B) as the selective marker. Both genes were inserted behind the constitutive *Aspergillus nidulans Gpd* promotor. The transformants were mitotically stabilized and insertion of the genes into the chromosomal DNA verified. In order to find one isolate that phenotypically resambled the wild type, three transformed isolates were examined *in vitro* for their morphological and physiological similarities to the wild type.

4. Population studies

One transformant T3a that in all tests resambled the wild type was then tested for its fitness in the rhizosphere of cucumber plants grown in a steamed sphagnum peat [10]. Population studies of the transformant (T3a) and the wild type (T3), expressed as cfu per gram sphagnum peat, showed that the transformant grew and increased in the rhizosphere just like the wild type. This was a final confirmation of the results from Thrane et al. [21], which showed that strain T3a phenotypically resembles the wild type (T3). Population development of the transformant is shown in figure 1.

By wounding the mycelium of the colonies with a cork borer and then adding the synthetic substrate (X-gluc), a test for GUS reaction could be carried out directly on the plates after dilution plating. In a treatment with a mixture of the transformant and the wild type, the proportion of colonies originating from the transformant was constant throughout the population studies. This indicated that the transformant was genetically stable when growing in a natural potting mixture [10].

The dilution plating technique has probably been one of the most used methods for determining populations of microorganisms in soil, and biomass are often derived from the counts obtained. The method has been much discredited, one of the major criticisms being that the majority of colonies arise from conidia rather than hyphae [23]. This is strongly supported by our results [10], which showed that an increase in cfu correlated to formation of conidia observed by FDA staining of peat samples. As such, the propagules which are counted are not necessarily those which are functioning in the ecosystem [3]. In limited circumstances, such as comparative studies of the GUS transformant and wild type strains on selective media or determination of e.g. inoculum density, dilution plating is useful. But when studying ecology, population dynamics, and function of microorganisms in soil and rhizosphere the major requirements are to measure growth rates and metabolic activities of individual species in the population rather than those of the total population itself.

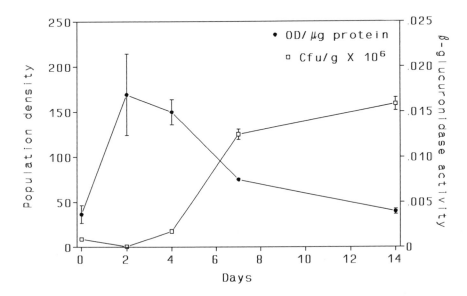

Figure 1. Population development of the GUS transformed strain of *T. harzianum*, expressed as cfu per gram dried sphagnum peat, and the corresponding ß-glucuronidase activity expressed as OD per μg extracted protein. Standard error is shown.

5. Biomass and activity

Many microbial ecologists have considered chemical determination an attractive possibility for biomass estimation, but most of the substances chosen as biomass indicators do not distinguish single species from the total microbial population. However, the integration of marker genes into the genome of microorganisms enables the production of compounds with the necessary degree of specificity. There have been several reports of the application of the GUS marker technique for biomass quantification of plant pathogens in infected plant tissue. Oliver et al. [14] used a GUS transformed strain of *Cladosporium fulvum* to distinguish between compatible and incompatible reactions in tomato host plants. Liljeroth et al. [13] found a positive correlation between GUS activity and lesion size in barley roots infected with a transformed strain of *Bipolaris sorokiniana*. Furthermore, they showed a positive correlation between the GUS activity and the content of the fungal biomass indicator, ergosterol, in infected host tissue. Couteaudier et al. [5] found a reduction of GUS activity in flax roots, exposed to a GUS+ pathogenic strain of *Fusarium oxysporum* in association with a nonpathogenic strain, which they interpret as the ability of the nonpathogenic strain to inhibit successful colonization by the pathogenic strain. However, quanitiative measurements of fungal biomass should always be interpreted with care, as the content of any biomass indicator can vary under different conditions [22].

We have measured the activity of our GUS transformed strain of *T. harzianum* by protein extraction and quantification of ß-glucuronidase activity from steamed sphagnum peat [10]. The ß-glucuronidase activity, expressed as optical density (OD) per microgram protein, increased dramatically the first few days of the experiment. After having reached a maximum after 2 days, it decreased until it, at day 14, reached the same level as day 0 (figure 1).

Comparison between results of the activity and population studies show that the two curves were inversively related. A possible explanation of this is that ß-glucuronidase production was high when the added conidia were preparing for germination and during the period of mycelia growth. After a few days the nutrients released by steaming of the peat is used up by the organism (Heiberg et al., this volume) causing an decrease in metabolic activity including ß-glucuronidase production. At this point mycelia growth stops accompanied by formation of conidia (i.e. increase in cfu).

This hypothesis is strongly supported by the results of an investigation of the actual growth phases of the organism [10], studied by FDA staining [20] of conidia and hyphae in peat samples. The number of swollen (active) and germinated conidia were highest (80 %) after two days. At day four, when formation of new conidia had started, the percentage of swollen and germinated conidia had decreased to 63 %. After 14 days less than 1 % of the conidia were swollen or germinated. Although scattered swelling, germination, and formation of conidia seemed to take place at any time thereafter, this had no detectable influence on the cfu and the ß-glucuronidase activity. Rather, it indicated a constant turnover of the total biomass. At any time during the experiment, actual biomass estimation based on measuring of hyphal lengths was impossible due to the scattered appearance of hyphal fragments. Formation of new conidia seemed to take place only a short time after germination, not leaving much time for hyphal growth.

In our as well as in the referred examples, production of ß-glucuronidase is controled by a constitutive promotor which makes probable that the production of the enzyme correlates with the general physiological condition of the organisms varying with different phases of their lifecycles and under influence of different environmental conditions. Therefore, the level of enzyme production rather than being an indication of the biomass qualify as an expression of the activity of the fungi. This is strongly supported by the fact that the ß-glucuronidase activity in peat samples correlates nicely with the proportion of swollen and germinated conidia (figure 2). In accordance with this, Eparvier and Alabouvette [8] considered a constitutively expressed GUS gene in *Fusarium oxysporum* f. sp. *lini* to reflect the metabolic activity of the strain. Thus, they found that the color intensity of X-gluc stained hyphae indicated the activity of the fungus on flax roots. Jefferson [11] and Wilson et al. [24] found that bacteria, fungi, and plants tested had, with some exceptions, only negligible background GUS activity. This makes it possible to study fungal activity in complex systems by extraction of the ß-glucuronidase, but one should be aware that in non sterile soil, background activity may occur due to the presence of bacteria or fungi with intrinsic GUS activity (Persmark et al., this volume).

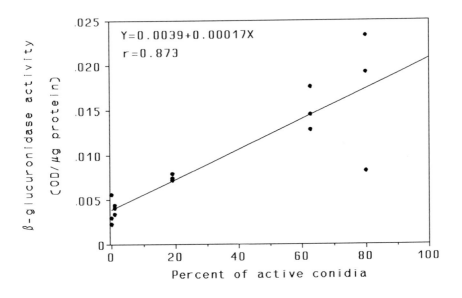

Figure 2. Correlation of the ß-glucuronidase activity and the percentage of active (swollen and germinated) conidia in peat samples taken at the same days [10].

6. Localization of hyphae on the rhizoplane

A methodology that enables the study of the GUS transformant directly on the rhizoplane has been developed [10]. The root systems of cucumber plants grown in sphagnum peat infested with the organism were incubated in the synthetic substrate (X-gluc). Blue stained hyphae could easily be seen under a stereomicroscope. No staining was seen on root systems from control plants grown in non infested peat or peat infested with the wild type. The occurance of stained hyphae varied from almost no hyphae on the total root systems to many small colonies of mycelia randomly spread over different regions of the roots. There did not seem to be preference for the root tip or any other region. Histological examination revealed that the hyphae were restricted to from one to a few neighboring cells mainly associated with small wounds such as damaged root hairs and other epidermal cells. These observations support the hypothesis that this ability is important for the antagonistic effect of *T. harzianum* as competition for nutrients leaking from the cells and actual colonization of the wound may block a possible way for pathogens to infect the plant.

7. References

1.   Abd-El Moity T.H., Papavizas, G.C., and Shatla, M.N. (1982) Introduction of new isolates of *Trichoderma harzianum* tolerant to fungicides and their experimental use for control of white rot of onion. Phytopathology 72:396-400.
2.   Ahmad, J.S., and Baker, R. (1987) Competetive saprophytic ability and cellulytic activity of rhizosphere competent mutants of *Trichoderma harzianum*. Phytopathology 77:358-362.
3.   Brock, T.D. (1987) The study of microorganisms *in situ*: progress and problems. Pages 1-17 in: Ecology of Microbial Communities. M. Fletcher, T.R.G. Gray and J.G. Jones, eds. Symp. Soc. Gen. Microbiol. Vol. 41.
4.   Chet, I. (1987) *Trichoderma*-application, mode of action, and potential as a biocontrol agent of soil-borne plant pathogenic fungi. Pages 137-160 in: Innovative Approaches to Plant Disease Control. I. Chet, ed. Wiley, New York.
5.   Couteaudier, Y., Daboussi, M.-J., Eparvier, A., Langin, T., and Orcival, J. (1993) The GUS gene fusion system (*Escherichia coli* ß-D-glucuronidase gene), a useful tool in studies of root colonization by *Fusarium oxysporum*. Appl. Environ. Microbiol. 59:1767-1773.
6.   Davet, P. 1979. Technique pour l´analyse des populations de *Trichoderma* et de *Gliocladium virens* dans le sol. Ann. Phytopathol. 11:529-533.
7.   Elad, Y., Chet, I., and Henis, Y. (1981) A selective medium for improving quantitative isolation of *Trichoderma* spp. from soil. Phytoparasitica 9:59-67.
8.   Eparvier, A., and Alabouvette, C. (1994) Competition between pathogenic and non pathogenic *Fusarium oxysporum* for root colonization. Biocontrol Sci. Technol. 4:35-47.
9.   Gallagher, S.R. (1992) GUS protocols: Using the GUS gene as a reporter of gene expression. Academic Press, San Diego, 221 pp.
10. Green, H., and Jensen, D.F. (1995) A tool for monitoring *Trichoderma harzianum:* II. The use of a GUS transformant for ecological studies in the rhizosphere. Phytopathology, 85:1436-1440.
11. Jefferson, R.A. (1987) Assaying chimeric genes in plants: the GUS gene fusion system. Plant Mol. Biol. Rep. 5:387-405.

12. Jensen, D.F., and Wolffhechel, H. (1995) The use of fungi, particularly *Trichoderma* spp. and *Gliocladium* spp., to control root rot and damping-off diseases. Pages 177-189 in: Biocontrol Agents: Benefits and Risks. H. Hokkanen, and J.M. Lynch eds. Cambridge University Press, Cambridge.

13. Liljeroth, E., Jansson, H.-B., and Schäfer, W. (1994) Transformation of *Bipolaris sorokiniana* with the GUS-gene and use for studying fungal colonization of barley roots. Phytopathology 84:1484-89.

14. Oliver, R.P., Farman, M.L., Jones, J.D.G., and Hammond-Kosack, K.E. (1993) Use of fungal transformants expressing ß-glucuronidase activity to detect infection and measure hyphal biomass in infected plant tissues. Mol. Plant-Microbe Interact. 6:521-525.

15. Papavizas, G.C. (1985) *Trichoderma* and *Gliocladium*: biology, ecology, and potential for biocontrol. Annu. Rev. Phytopathol. 23:23-54.

16. Papavizas, G.C., Lewis, J.A., and Abd-El Moity, T.H. (1982) Evaluation of new biotypes of *Trichoderma harzianum* for tolerance to benomyl and enhanced biocontrol capabilities. Phytopathology 72:126-132.

17. Papavizas, G.C., and Lumsden, R.D. (1982) Improved medium for isolation of *Trichoderma* spp. from soil. Plant Dis. 66:1019-1020.

18. Parkinson, D., and Coleman, D.C. (1991) Methods for assessing soil microbial populations, activity and biomass. Agric. Ecosyst. Environ. 34:3-33.

19. Sambrook, J., Fritsch, E.F., and Maniatis, T. (1989) Molecular cloning. A laboratory manual. Second edition. Cold Spring Harbour Laboratory Press, Cold Spring Harbour.

20. Söderström, B.E. (1977) Vital staining of fungi in pure cultures and in soil with fluorescein diacetate. Soil Biol. Biochem. 9:59-63.

21. Thrane, C., Lübeck, M., Green, H., Defégu, Y., Allerup, S., Thrane, U., and Jensen, D.F. (1995) A tool for monitoring *Trichoderma harzianum*: I. Transformation with the GUS gene by protoplast technology. Phytopathology, 85:1428-1435.

22. Torres, M., Viladrich, R. Sanchis, V., and Canela, R. (1992) Influenxe of age on ergosterol content in mycelium of *Aspergillus ochraceus*. Lett. Appl. Microbiol. 15:20-22.

23. Warcup. J.H. (1955) On the origin of colonies of fungi developing on soil dilution plates. Trans. Br. Mycol. Soc. 38:298-262.

24. Wilson, K.J., Giller, K.E., and Jefferson, R.A. (1991) ß-glucuronidase (GUS) operon fusion as a tool for studying plant-microbe interactions. Pages 226-229 in: Advances in Molecular Genetics of Plant-Microbe Interactions. Vol. 1. H. Hennecke, and D.P.S. Verma, eds. Kluwer Academic Publishers, the Netherlands.

Monitoring growth of *Bipolaris sorokiniana* in plant tissue using GUS (β-glucuronidase) as a marker.

E. LILJEROTH
Department of Plant Breeding Research, The Swedish University of
Agricultural Sciences, S-268 31 Svalöv, Sweden

Introduction
The phytotoxin producing fungus *Bipolaris sorokiniana* is a severe pathogen on barley and other cereals and may infect above ground as well as below ground tissues of the host plant. Host resistance to *B. sorokiniana* varies quantitatively among different cultivars. No total resistance occurs but it appears that relatively resistant cultivars can resist fungal development better within the host tissue, but not stop fungal penetration. Ecological studies on the infection of plants, especially in the rhizosphere are few, mainly due to a lack of suitable detection and quantification methods. Biomarkers, e.g. ergosterol, have been used for monitoring unspecific fungi. For studies of a particular fungus a more specific marker would be desirable. Methods to transform fungi offer possibilities to introduce markers that can later be monitored. I have successfully transformed a marker gene (GUS, β-glucuronidase from E. coli) into a strain of *B. sorokiniana* (Liljeroth *et al.*, 1993) which makes it possible to quantify fungal growth in infected plant tissue. However, to be able to get reliable results from experimental studies, knowledge of the stability of the transformant, the variation in specific expression of the marker gene and the correlation with other recognized methods is necessary. This paper deals with these questions and also with measurements of fungal growth in root tissue of barley and other cereals in relation to root characters such as resistance and root cortical cell death.

Methods
The technique for transformation of *B. sorokiniana* is described in Liljeroth *et al.* (1993). A short description of the experimental methods is presented here. For details see Liljeroth *et al.* (1993), Liljeroth (1995) and Liljeroth *et al.* (submitted).

One transformant strain was used for the experiments. The ability of the transformant to compete with the wild type was studied after synchronous inoculation in sterilized soil and on oat kernels. Erlenmeyer flasks containing 100 gram soil each, were inoculated with 300 conidia per gram of soil of both the wild type and the transformant. The numbers of CFU of both strains were determined at specific times.

The relation between GUS-expression, hyphal dry weight and protein content was

D. F. Jensen et al. (eds.), Monitoring Antagonistic Fungi Deliberately Released into the Environment, 85–89.

studied in pure culture. Erlenmeyer flasks with 30 ml Czapec Dox liquid medium were inoculated with 5000 conidia of the fungus and incubated on a rotary shaker at 200 rpm, 25 C. At specific time points hyphal dry weight, protein content and GUS activity were determined in three replicate erlenmeyers. The GUS-activity was determined fluorimetrically using MUG (4-methyl-umbelliferyl-β-D-glucuronide) as substrate (Jefferson, 1987).

Roots of 10 day old cereal seedlings grown on filter paper were inoculated with an agar disc containing 1000 conidia of the transformed fungus (see Fig. 1). The paper was rolled together and the lower end placed in distilled water. At specific time points 6 replicate paper rolls were opened and the size of the lesions were measured. Proteins were extracted from the roots, excluding the inoculation disc, and total GUS-activity in the extract was determined as described above.

Root cortical cell death (RCD) was determined by staining root segments of 25 day old plants with acridine orange and assessing the number of viable nuclei in the cortical tissue with fluorescence microscopy (Henry and Deacon, 1981).

Results and Discussion

The transformed strain of *B. sorokiniana* was found to be stable, i.e. it expressed the GUS-gene in a quantitatively similar manner after several conidiation cycles (Liljeroth *et al.*, 1993). The ability of the transformant to compete with the wildtype was studied after simultaneous inoculation of soil with conidia from the transformant and the wild type. None of the strains outcompeted each other and the frequency of colony forming units consisting of transformant decreased only slightly during the first months of incubation (Figure 1). This result indicated that the competitive ability of the transformant was not very much changed and that it is likely that the behaviour of the transformant is similar to that of the wild type at least over the time period of most laboratory experiments.

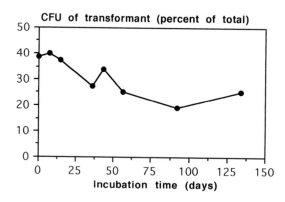

Figure 1. Change in the frequency of GUS-transformed CFU in soil. At start equal numbers of wild type and transformant conidia were inoculated into the soil

The fungal development in plant tissue was analysed by measuring the activity of the GUS enzyme after protein extraction of the tissue. These values were found to be significantly correlated with ergosterol content in the plant tissue (Liljeroth *et al.*, 1993). Since ergosterol content is a recognized method for measuring fungal biomass and has been used for fungal growth assessment in plant tissue (Gordon and Webster, 1984; Newell *et al.*, 1988) GUS-activity appears to correlate well with fungal biomass. In pure culture experiments the GUS-activity of hyphal extracts correlated well with hyphal dry weight and with total protein content (Figure 2)

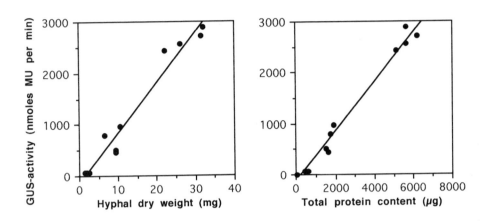

Figure 2. Correlation between GUS-activity and hyphal dry weight/total protein content in pure culture of a GUS transformed strain of *B. sorokiniana*.

After these basic but important investigations the transformant was used to study fungal growth in roots of different cereals. Roots of 10-day old seedlings grown on filter paper were inoculated with the fungus and at different time intervals lesion size and GUS-activity in the roots were determined. The fungus had a slower growth rate in resistant barley varieties, i.e. varieties producing smaller lesions, than in relatively susceptible varieties (Liljeroth *et al.*, submitted). This is not surprising but also not self-evident since cultivars could respond differently to infection by producing lesions of different sizes. With this method it is possible to differentiate cultivars which can restrict the fungal invasion of the plant tissue. De la Pena and Murray (1994) used the GUS-method for studying infection by *Pseudocercosporella herpotrichoides* in wheat. They reported that this method enabled better differentiation among cultivar resistance and susceptibility than visual disease ratings.

The fungal root colonization was further investigated with different cereal species. The development of both lesion size and GUS-activity varied significantly in roots of different cereal species after inoculation with GUS-transformed *B. sorokiniana*. The species may have different biochemical defense mechanisms, but other physiological differences can also be important. Liljeroth (1995) studied the rate of senescence in cortex cells (RCD) in uninfected seedlings of cereals. The fraction of the cortex containing viable cells differed greatly among the investigated cereals species. It appeared that species with a relatively fast rate of cortical senescence had higher levels of GUS-activity, i.e. larger fungal biomass, in roots after inoculation with the transformed fungus (Figure 3a). This result indicates that cortical senescence is important for the fungal establishment in the roots.

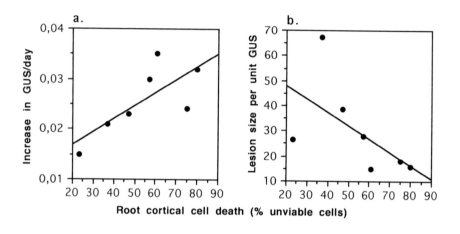

Figure 3 a. Relation between root cortical cell death (RCD) in uninfected roots and fungal biomass increase, measured as GUS-activity, after inoculation with transformed *B. sorokiniana*. b. Relation between RCD and lesion size development per unit fungal biomass. The investigated cereals species were, from left to right, in both figures: rye, barley, *Triticum monococcum*, *T. dicoccum*, triticale, *T. spelta*, wheat.

In wheat and triticale, fungal growth was faster than in barley despite the fact that the lesions were smaller and lighter in these species. The fungal root invasion in different *Triticum* species was also investigated. There were clear differences in lesion size development per unit fungal biomass among the cereal species (Figure 3b). Species that produced the largest and most coloured lesions per unit fungal biomass present in the roots after inoculation, had a lower degree of root cortical death in uninfected roots, i.e. a larger fraction of the cortex had viable nuclei (Liljeroth, 1995). Only viable cortex cells can respond to infection by producing compounds that cause browning of the root tissue. This might explain why wheat and triticale produced the smallest and least coloured lesions even though the roots of those species were most heavily infected.

Concluding remarks

Transforming fungi with marker genes offers interesting possibilities for ecological studies of single fungal strains. The growth and invasion of of fungi in plant tissue can be carefully investigated in relation to defense mechanisms of the host plant which was shown for *B. sorokiniana* here and for *Pseudocercosporella herpotrichoides* by De la Pena et al. (1994). Advantages of the method are improved sensitivity and specificity and that the procedure for extraction and GUS-assay is simple. With a system for automatic sampling, many samples could be run in a short time which is essential for these kinds of experiments. However, regulations governing the release of genetically altered organisms limits its use under field conditions. An important application of this technique is to monitor growth and survival of biocontrol fungi in a manner more precise than counting colony forming units. This has been suggested and illustrated by Couteaudier *et al.* (1993) and Epavier and Alabouvette (1994).

References

Couteaudier, Y., Daboussi, M.-J., Eparvier, A., Langin, T., and Orcival, J. 1993. The GUS gene fusion system *(Escerichia coli* β-D-glucuronidase gene), a useful tool in studies of root colonization by Fusarium oxysporum. *Appl. Environ. Microbiol.* **59 : 6**, 1767-1773.

Eparvier, A. and Alabouvette, C. 1994. Use of ELISA and GUS-transformed strains to study competition between pathogenic and non-pathogenic *Fusarium oxysporum* for root colonization. *Biocontrol Science and Technology* **4**, 35-47.

Gordon, T.R., and Webster, R.K. 1984. Evaluation of ergosterol as an indicator of infestation of barley seed by *Drechslera graminea*. *Phytopathol.* **7 4**, 1125-1127.

De la Pena, R.C. and Murray, T.D. (1994) Identifying wheat genotypes resistant to eyespot disease with a b-glucuronidase.transformed strain of *Pseudocercosporella herpotrichoides*. *Phytopathol.* **8 4**, 972-977.

Henry, C.M. and Deacon, J.W. (1981) Natural (non-pathogenic) death of the cortex of wheat and barley seminal roots, as evidenced by nuclear staining with acridine orange. *Pl. Soil* **6 0**, 255-274.

Jefferson, R.A. (1987) Assaying chimeric genes in plants: The GUS gene fusion system. *Plant Mol. Bio. Rep.* **5**, 387-405.

Liljeroth, E., Jansson, H-B. and Schäfer, W. (1993) Transformation of *Bipolaris sorokiniana* with the GUS gene and use for studying fungal colonization of barley roots. *Phytopathol.* **8 3**, 1484-1489.

Liljeroth, E. (1995) Comparisons of early root cortical senescence among barley cultivars, Triticum species and other cereals. *The New Phytologist* **1 3 0**, 495-501.

Liljeroth, E., Franzon-Almgren, I. and Gunnarsson, T. Root colonization by Bipolaris sorokiniana in different cereals and relations to lesion development and natural root cortical cell death. *J. Phytopathol.* (submitted)

Newell, S.Y., Arsuffi, T.L. and Fallon, R.D. 1988 Fundamental procedures for determining ergosterol content of decaying plant material by liquid chromatography. *Appl. Environ. Microbiol.* **54 : 7**, 1876-1879.

IDENTIFICATION OF FUNGI BY SECONDARY METABOLITES

ULF THRANE
Mycology Group
Department of Biotechnology
Technical University of Denmark
DK-2800 Lyngby
Denmark

1. Introduction

Classical procedures for identification of fungi at species level are based on observation of morphological characters, in some cases supported by visual detection of pigmentation. For mushrooms and other fleshy fungi tasting and sniffing to the specimen are also used. The fungal taxonomy which is the base for identification keys is mainly based on morphological characters. However, morphology is only part of the extrovert differentiation. Another part is a chemical fingerprint of biological active compounds such as antibiotics, exoenzymes, volatiles, toxic compounds, pigments a.o. The chemical characters should be regarded just as important in taxonomy as morphological characters. Similar to morphological structures production of secondary metabolites are based on expressions of hundreds of genes. The profile of biological active chemicals is very important for survival of an organism in nature. The need of multidisciplinary data within fungal taxonomy is now accepted among many mycologists which is reflected by an increasing number of publications on fungal systematics using a holistic approach.

D. F. Jensen et al. (eds.), Monitoring Antagonistic Fungi Deliberately Released into the Environment, 91–98.
© 1996 *Kluwer Academic Publishers. Printed in the Netherlands.*

The use of secondary metabolites for fungal classification has been surveyed for many years. Within the last 20 years the analytical chemistry has improved dramatically and a lot of chemical data has been obtained. At this laboratory emphasis has been given to studies within the genera *Penicillium, Aspergillus, Fusarium,* and *Alternaria.* The studies have focused on the profile of non-volatile secondary metabolites; however, recently profiles of volatiles have been included as well.

Many secondary metabolites are produced by a range of different fungal species. However, a profile of two or more specific compounds has a limited distribution. The more compounds in the profile the higher the specificity. By five metabolites (different biosynthetic pathways) it is possible to segregate even closely related species.

2. Non-volatile secondary metabolites

2.1. THIN LAYER CHROMATOGRAPHY

The profile of non-volatile secondary metabolites of pure fungal cultures can be determined by a fast and simple thin layer chromatographic (TLC) method. A small (4 mm in diameter) agar plug is cut from the agar culture by a cork borer. The plug is then applied onto the TLC plate, either placing the agar or the mycelium onto the plate. In the latter case intracellular metabolites are extracted by wetting the mycelium with a drop of chloroform-methanol prior to application. After removal of the plugs, the TLC plate is developed by conventional methods. More details are given by Filtenborg *et al.* (1995). Within 7 to 14 days of incubation most fungal species produce many secondary metabolites which can be visualized as a specific pattern of coloured spots on the developed TLC plate. By comparison to metabolite standards many spots can be identified; however, many unknown spots are present. A TLC UV-Scanner may be used to verify the identity of the spots by their UV-reflectance spectrum compared to a spectral library. In practice, identification of a given isolate may not need exact identification of one or more TLC spots, as the full TLC pattern of coloured spots in itself is species specific. In this case TLC patterns of well characterized and identified reference strains are needed. At this point it needs to be emphasized that the reference strains should be identified by an authority before use and standardized cultural conditions should be followed.

For screening purposes and grouping isolates into homogenous units the agar plug TLC method is very useful. It is an easy, fast and cheap method which is very powerful for taxonomic work as well as for routine identification.

2.2. HIGH PERFORMANCE LIQUID CHROMATOGRAPHY[*]

More detailed information on the secondary metabolites can be obtained by analysis of fungal extracts by high performance liquid chromatography with diode array detection (HPLC-DAD). Agar cultures are extracted by organic solvents, concentrated and defatted before HPLC-DAD analysis. More details can be found in a recent review by Frisvad & Thrane (1993). By HPLC-DAD the UV spectrum (200-600 nm) from each separated compound is recorded and stored in the computer system. A 3-dimensional plot of a chromatogram including the UV spectra is shown in Figure 1. Unattended analyses of many samples, e.g. overnight, including data analysis and print-outs of the results are possible.

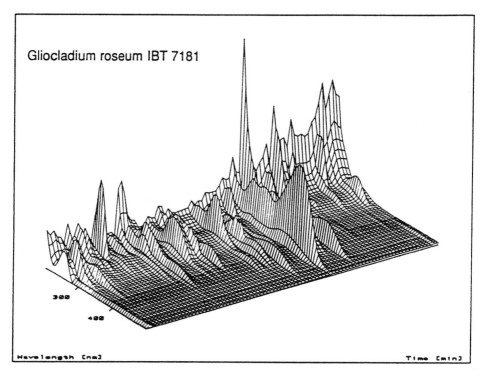

Figure 1. A plot of a HPLC-DAD chromatogram of an analysis of a culture extract of *Gliocladium roseum* IBT 7181.

To overcome fluctuations in retention time and thus problems in comparing results from different analyses, a set of alkylphenone standards is analysed for each 15 samples. The retention times from analysis of the alkylphenones are used for calculating a retention time index for each peak in the following 15 samples. The retention index is stable to new batches of solvents, columns etc (Frisvad & Thrane 1987; 1993).

By detailed data analysis and comparison to UV spectral libraries several peaks in the chromatograms may be identified. In the chromatogram in Figure 1 no peaks could be identified; however, unidentified HPLC peaks (Figure 2) can be used for characterizing and grouping fungal isolates using the full HPLC profile of known and unknown compounds.

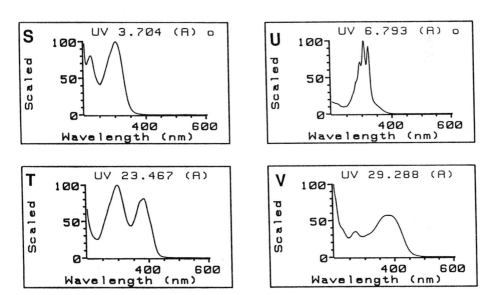

Figure 2. UV spectra (200 - 600 nm) of four peaks of the chromatogram of *Gliocladium roseum* IBT 7181 shown in Figure 1.

3. Volatile secondary metabolites

For ages mycologists have been sniffing to their specimens. However, correlation between the profile of volatiles and fungal systematics has been given little attention so far. A study focussing on detection and identification of volatile metabolites from *Penicillium* species have been performed by Larsen & Frisvad (1995a; 1995b). By simple methods for collection of volatile metabolites (Larsen & Frisvad, 1994) they have found that each fungal species has its own specific profile of volatiles. Using sensitive chemical analysis by gaschromatography with mass spectrometric detection (GC-MS) fungal growth could be detected within few days of incubation. As the sampling methods are non-destructive GC-MS may easily be applied together with conventional methods for examining fungal growth.

4. Identification keys

To construct a reliable identification key a sound taxonomy based on multidisciplinary data is needed. One way to go is to combine chemical, physiological and morphological data. By combining chemical data from the different chromatographic analyses each fungal species has its own profile of secondary metabolites. During the last 20 years data have been compiled extensively at this laboratory (Frisvad *et al.*, 1989; Filtenborg *et al.*, 1995). Much work have been done within the genera *Aspergillus, Penicillium, Fusarium, Alternaria,* and recently *Trichoderma.* Unfortunately, a lot of misidentified producerstrains or misidentified metaboliteproduction (or both!) is reported in the international literature. Updated information trying to clarify the misinformation have been published for *Aspergillus* and *Penicillium* (Frisvad, 1989) and *Fusarium* (Thrane, 1989), but the new updates should be published soon.. Recently an update focussing on mycotoxins has been published (Frisvad & Thrane, 1995).

Identification keys based on a combined use of profiles of secondary metabolites, morphological characters, and simple physiological data have been generated for common species within *Penicillium, Aspergillus, Fusarium,* and *Alternaria.* Most keys have been printed analytical or synoptical keys, however, for *Fusarium* species a simple computerized key, FUSKEY, has been constructed (Thrane, 1991). FUSKEY is based on public software packages: DELTA (DEscription Language for TAxonomy) and INTKEY (INTeractive program for KEYing specimens) which are available via Internet. Computerized keys for *Penicillium* species and *Alternaria* species based on neural network technology are in progress. These keys use multidisciplinary data and seem very promising.

There is no doubt that future keys will be computerized keys like the exciting interactive keys for birds, fishes, sea cucumbers, butterflies, grasses, rain forest trees a.o. which are available on CD-ROM or via Internet.

5. References

Filtenborg, O., Frisvad, J.C., Thrane, U., and Lund, F. (1995) Screening methods for secondary metabolites produced by fungi in pure culture, in R.A. Samson, E.S. Hoekstra, J.C. Frisvad, and O. Filtenborg (eds.), *Introduction to food-borne fungi*, Centraalbureau voor Schimmelcultures, Baarn, pp. 270-274.

98

Frisvad, J.C. (1989) The connection between the penicillia and aspergilli and mycotoxins with special emphasis on misidentified isolates, *Arch. Environ. Contam. Toxicol.* **18**, 452-467.

Frisvad, J.C. and Thrane, U. (1987) Standardized high-performance liquid chromatography of 182 mycotoxins and other fungal metabolites based on alkylphenone indices and UV-VIS spectra (diode-array detection), *J. Chromatogr.* **404**, 195-214.

Frisvad, J.C. and Thrane, U. (1993) Liquid column chromatography of mycotoxins, in V. Betina (ed.), *Chromatography of mycotoxins: techniques and applications*, Elsevier Publishers B.V., Amsterdam, pp. 253-372.

Frisvad, J.C. and Thrane, U. (1995) Mycotoxin production by food-borne fungi, in R.A. Samson, E.S. Hoekstra, J.C. Frisvad and, O. Filtenborg (eds.), *Introduction to food-borne fungi*, Centraalbureau voor Schimmelcultures, Baarn, pp. 251-260.

Frisvad, J.C., Filtenborg, O., and Thrane, U. (1989) Analysis and screening for mycotoxins and other secondary metabolites in fungal cultures by thin-layer chromatography and high-performance liquid chromatography, *Arch. Environ. Contam. Toxicol.* **18**, 331-335.

Larsen, T.O. and Frisvad, J.C. (1994) A simple method for collection of volatile metabolites from fungi based on diffusive sampling from Petri dishes, *J. Microbiol. Meth.* **19**, 297-305.

Larsen, T.O. and Frisvad, J.C. (1995a) Chemosystematics of fungi in the genus *Penicillium* based on profiles of volatile metabolites, *Mycol. Res.* **99**, 1167-1174.

Larsen, T.O. and Frisvad, J.C. (1995b) Characterizatrion of volatile metabolites from 47 taxa in genus *Penicillium. Mycol. Res.* **99**, 1153-1166.

Thrane, U. (1989) *Fusarium* species and their specific profiles of secondary metabolites, in J. Chelkowski (ed.), Fusarium: *Mycotoxins, taxonomy and pathogenicity*, Elsevier Science Publishers B.V., Amsterdam, pp. 199-225.

Thrane, U. (1991) FUSKEY, an interactive computer key to common *Fusarium* species, *Mycotoxin Res.* **7A**, 50-53.

PREHELMINTHOSPOROL, A PHYTOTOXIN FROM *BIPOLARIS SOROKINIANA*

H. ÅKESSON AND H.-B. JANSSON
Department of Microbial Ecology, Lund University
Ecology Building, 223 62 Lund, Sweden

1. Introduction

Bipolaris sorokiniana (Sacc. in Sorok.) Shoem. (syn. *Helminthosporium sativum* (Pamm. King & Backe), teleomorph *Cochliobolus sativus* (Ito & Kurib.) is a widespread fungus which can cause disease in grasses including cereals but occasionally also other taxonomic groups (Wildermuth and MacNamara, 1987). It attacks all parts of the plants from roots to seeds, at all stages of plant growth. Depending on the site of infection, it causes diseases such as spot blotch, seedling blight, foot rot, crown rot and root rot on cereals. The fungus is transmitted by seeds, is soil borne or is dispersed as conidia by air currents. The diseases occur mainly in the warmer cereal growing regions, particularly in North America and parts of Australia and New Zealand (Sivanesan and Holliday, 1981). They were regarded relative unimportant in Europe (Sivanesan and Holliday, 1981), but have become increasingly important in barley in the cool climate of North-Western Europe (Jørgensen, 1974; Kurppa, 1984).

Toxins are thought to be important in the infection process. *B. sorokiniana* is known to produce a series of sesquiterpenoid metabolites such as prehelminthosporol, prehelminthosporolactone, helminthosporal, helminthosporol and victoxinine. Recent studies proposed prehelminthosporol (Fig. 1) to be the dominant sesquiterpene produced by the fungus and the most active in disruption of plant membranes (Carlson, *et al.* 1991a).

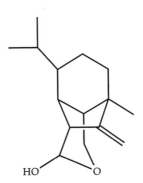

Fig. 1. Chemical structure of prehelminthosporol ($C_{15}H_{24}O_2$)

99

D. F. Jensen et al. (eds.), Monitoring Antagonistic Fungi Deliberately Released into the Environment, 99–104.
© 1996 *Kluwer Academic Publishers. Printed in the Netherlands.*

2. Prehelminthosporol

2.1. PRODUCTION *IN VITRO*

Prehelminthosporol has been isolated from culture filtrates of liquid cultures of *B. sorokiniana* (Aldridge and Turner, 1970; Cutler *et al.*, 1982; Pena-Rodrigues *et al.*, 1988, Carlson *et al.*, 1991a). Large amounts of toxin has also been detected in the mycelium and in conidia collected from agar colonies (Carlson *et al.*, 1991a, Nilsson *et al.*, 1993).

Prehelminthosporol can be quantified by several chromatographic techniques: high preformance liquid chromathography (HPLC) using refractive index-detection, on-line HPLC-mass spectrometry (MS) and gas chromathography (GC) (Carlson *et al.* 1991a, Nilsson *et al.*, 1993). The GC method is a sensitive and the most rapid method and can even be performed without derivatization of the toxin (Nilsson *et al.*, 1993). Using this method it is possible to detect prehelminthosporol in a crude extract from approximately 700 conidia (Nilsson *et al.*, 1993). This fact suggests that it might be possible to use as a marker for *B. sorokiniana* in plants. However, the content of prehelminthosporol in fungal tissues varies with age (Nilsson *et al.*, 1993) and between different isolates of the fungus (Åkesson, Apoga, Odham and Jansson, unpublished). Therefore it could be used for detection of the fungus rather than a biomass marker. Prehelminthosporol is not only produced by *B. sorokiniana* but also by *B. zeicola, B. victoriae and Drechslera setariae* . However, it is not present in all *Bipolaris* and *Drechslera* species; it does not occur in *B. sacchari, B. oryzae, B. maydis, D. avenae, D. graminea* and *D. triticirepentis* (Åkesson, Apoga, Odham and Jansson, unpublished).

2.2. TOXIC EFFECTS ON PLANTS

Prehelminthosporol is a non-host selective toxin and interacts with cellular membranes of non-host plants (Carlson *et al.*, 1991a) as well as host plants. Liljeroth *et al.* (1994) showed that the toxin caused increased leakage of ATP from intact roots of barley seedlings. Membrane interactions can also be observed as reduced levels of proton pumping, ATP hydrolysis and calcium uptake in prehelminthosporol treated plasma membrane vesicles of barley roots (Olbe *et al.* 1995). Other toxic effects seen on barley root cells include an increased rate of nuclear disintegration in the cortex (Liljeroth *et al.*, 1994). Olbe *et al.* (1995) showed that callose synthase located in the plasma membrane was stimulated at low concentrations of prehelminthosporol, while it was inhibited at higher concentrations. All these physiological studies support the hypothesis that the degeneration of uninfected host cells as well as the host responses (papillae) seen in ultrastructural studies of barley roots infected by *B. sorokiniana* (Carlson *et al.*, 1991b) might be caused by prehelminthosporol.

Prehelminthosporol causes lesions similar in appearance to those induced by the fungus in the field, when applied to leaves of host plants (Pena-Rodriguez *et al.*, 1988). Cutler *et al.* (1982) showed inhibition of wheat coleoptile growth at concentrations of 10^{-3} and 10^{-4} M while lower concentrations (10^{-5} M) promoted growth. In addition, prehelminthosporol appears to be involved in softening or dissolution of the wax layer of the leaves. Prehelminthosporol has amphiphilic properties which suggest that it may function as a detergent and aid in softening of the wax. The fact that prehelminthosporol has been detected in hyphal sheaths of *B. sorokiniana* and that leaves treated with prehelminthosporol showed a disturbed wax layer, suggests that it may be important for softening or dissolution of the wax layer and thereby for adhesion of the fungus to the leaf surface (Åkesson, 1995).

2.3. ANTI-MICROBIAL EFFECTS

Non-selective toxins from plant pathogens often also affect other microorganisms (Panopoulos *et al.*, 1984). Anti-microbial activity has therefore sometimes been exploited in developing sensitive microbial assays for these toxins and in identification of their target systems (Panopoulos *et al.*, 1984). In some cases the toxin may play a role outside the pathogen-plant interactive phase, such as in normal cellular metabolism or in the saprophytic phase. Anti-microbial action may help the pathogen to minimize competition by soil-resident or plant-associated microflora. Tyner (1966) showed that *B. sorokiniana* competed poorly with other soil fungi in terms of its rate of growth but that extracts from the fungus inhibited growth of all fungi tested.

Prehelminthosporol is a phytotoxin that also affects microbial cells. In a screening of various microorganisms a toxic action on both bacteria and fungi was observed (Åkesson, 1995). Among the bacteria that were tested, only the Gram positive ones were affected by prehelminthosporol (Table 1). Anti-fungal activity was studied in a spore germination assay (Table 2). All fungi, including *B. sorokiniana* itself, showed inhibited spore germination in the presence of prehelminthosporol. The influence of the toxin on germination of *B. sorokiniana* spores was, however, comparatively weak.

At the highest concentration of prehelminthosporol the germ tubes of the germinated spores of *B. sorokiniana* were highly branched. Spores of the zygomycete *Rhizopus arrhizus* showed increased swelling in the presence of high concentrations of prehelminthosporol. The same effect was observed when the spores were allowed to grow in medium containing the membrane disrupting agent dimethyl sulfoxide

TABLE 1. Antibacterial activity of prehelminthosporol in an agar diffusion assay (from Åkesson, 1995)

	Bacteria	Antibacterial activity
Gram positive	*Bacillus megaterium*	+
	Bacillus subtilis	+
	Bacillus thuringiensis	+
	Staphylococcus epidermis	+
Gram negative	*Proteus vulgaris*	-
	Escherichia coli	-
	Shigella sonnei	-
	Salmonella typhimurium	-
	Pseudomonas aeruginosa	-
	Kleibsella pneumoniae	-

100 µg of prehelminthosporol dissolved in 20 µl of methylene chloride were added to filter paper disks (5 mm diameter). After evaporation of the solvent, the filter paper disks were put on agar plates (tryptone glucose extract agar (Difco) or Todd Hewitt agar (Difco), 3 disks/plate) seeded with the bacterium. After incubation at 37°C for 2 days the inhibition zone was measured. The bacteria that were regarded as inhibited (+) showed an inhibition zone of more than 1 mm. No inhibition was seen around control disks (evaporated solvent only).

TABLE 2. Antifungal activity of prehelminthosporol in a spore germination assay (from Åkesson, 1995)

Fungus	Inhibition of spore germination
Bipolaris sorokiniana [a, 1]	+
Gliocladium roseum [a, 3]	++
Microdochium bollei [a, 3]	+++
Arthrobotrys oligospora [a, 2]	+++
Rhizopus arrhizus [b, 4]	++++
Mortierella isabellina [b, 5]	+++++

About 500 spores were added to 100 µl of medium (glucose, 1.1 mM; KH_2PO_4, 1.8 mM; $MgSO_4 \times 7H_2O$, 5.1 mM; [a]KNO_3, 2.0 mM; [b]$(NH_4)_2SO_4$, 9.9 mM), containing different concentrations of prehelminthosporol (dissolved in 1% methanol), in microtitre wells. [1], incubated for 5 h at 20 °C; [2], 9 h at 25 °C; [3], 18 h at 20 °C; [4], 15 h at 20 °C; [5], 24 h at 20 °C. Spore germination at each concentration were compared to that of a control (only solvent added). +++++, less than 4 µg/ml needed for 50 % inhibition of spore germination; ++++, 4-15 µg/ml needed for 50 % inhibition; +++, 15-30 µg/ml; ++, 30-60 µg/ml; +, 60-120 µg/ml.

(DMSO). This indicates that the plasma membrane is also an important target of prehelminthosporol in microbial cells. Experiments with zoospores from the chytridomycete Catenaria anguillulae provide further evidence of this membrane effect. The zoospores lack a cell wall, the membrane is therefore directly exposed to the environment. Prehelminthosporol caused membrane disruption (as shown by leakage of ATP) and rapid death (viewed by light microscopy) of the zoospores (Åkesson, 1995).

2.4. MECHANISMS OF SELF-PROTECTION AGAINST THE TOXIN

In order to avoid lethal or deleterious effects of a non-selective toxin such as prehelminthosporol, the producer requires a system to protect itself. As B. sorokiniana seems to be somewhat less sensitive to the extracellular toxin than other microorganisms, the composition and properties of its plasma membrane may be different. Preliminary studies of membrane bound phospholipids showed a higher proportion of phosphatidyl inositol in B. sorokiniana compared to some other fungi (Torsten Gunnarsson, personal communication).

Internal concentrations of prehelminthosporol seem to be high in both conidia and hyphae of B. sorokiniana (Carlson et al., 1991a). How is the fungus able to store and use prehelminthosporol and at the same time be able to protect itself from deleterious effects of the toxin? To try to solve this problem antibodies against prehelminthosporol were prepared and used for subcellular localization of the toxin in B. sorokiniana using transmission electron microscopy (Åkesson, 1995). Several preparation methods were tried but fast-freeze fixation together with low-temperature preparation methods seemed to be most successful in localizing the toxin in conidia and hyphae. The fungal tissues in the methacrylate resin Lowicryl HM20 were treated according to methods for immunocytochemical (immunogold) labelling and examined in the transmission electron microscope. No labelling was seen in the cell wall and the vacuoles, but a weak labelling was found in the cytoplasm. However, the main labelling was confined to membrane-bound organelles identified as Woronin bodies. Woronin bodies have been described as spherical to ovoid single-membrane bound organelles located close to septal pores in ascomycete and deuteromycete fungi (Markham, 1994). They are also often seen in the hyphal apex, probably due to their synthesis there (Brenner and Carroll, 1968). Their main function is thought to be in occlusion of the septal pore in response to hyphal damage. In this way they can prevent excessive loss of cytoplasm which would otherwise result from hyphal rupture. Woronin bodies may also have other functions in the hyphae. The main component of these organelles is thought to be protein (McKeen, 1971; Mason and Crosse, 1975) and it has therefore been suggested that they may act as sites for storage of protein or nitrogen (Collinge and Markham, 1987; Jennings, 1989). Our results suggest that they may also function in the storage of hydrophobic compounds such as prehelminthosporol.

3. Conclusions

Prehelminthosporol is a sesquiterpene that has been isolated from *in vitro* cultures of *B. sorokiniana*. It is present in high amounts in hyphae and conidia and is also released into the surrounding medium. Prehelminthosporol is a non-selective toxin that acts on cells of both host and non-host plants. The main target seems to be the cell membranes but studies also indicate that the outer wax layer of aerial plant parts is dissolved or softened by toxin released from the fungus. Not only plant cells are affected by the toxin but also microbial cells including *B. sorokiniana* itself. Immunocytochemical localization of the toxin in cells of *B. sorokiniana* indicates that the fungus avoids deleterious effects of prehelminthosporol during production and storage by keeping it in the Woronin bodies.

The importance of the toxin in the infection process, as well as the possibilities to use prehelminthosporol as a biomarker for fungal infection is presently being investigated.

4. References

Åkesson, H. (1995) Infection of barley by *Bipolaris sorokiniana*: toxin production and ultrastructure. Ph. D. thesis. Department of Ecology. Lund University.

Aldridge, D. C. and Turner, W. B. (1970) 9-hydroxyprehelminthosporol, a metabolite of *Cochliobolus* (*Helminthosporium*) *sativus*. *J. Chem. Soc.* C, 686-688.

Brenner, D. M. and Carroll, G. C. (1968) Fine-structural correlates of growth in hyphae of *Ascodesmis sphaerospora*. *J. Bacteriol.* **95**, 658-671.

Carlson, H., Nilsson, P., Jansson, H.-B. and Odham, G. (1991a) Characterization and determination of prehelminthosporol, a toxin from the plant pathogenic fungus *Bipolaris sorokiniana*, using liquid chromatography/mass spectrometry. *J. Microbiol. Methods* **13**, 259-269.

Carlson, H., Stenram, U., Gustafsson, M. and Jansson, H.-B. (1991b) Electron microscopy of barley root infection by the fungal pathogen *Bipolaris sorokiniana*. *Can. J. Bot.* **12**, 2724-2731.

Collinge, A. J. and Markham, P. (1987) Nuclei plug septal pores in severed hyphae of *Sordaria brevicollis*. *FEMS Microbiol. Lett.* **44**, 85-90.

Cutler, H. G., Crumley, F. G., Cox, R. H., Davis, E. E., Harper, J. L., Cole, R. J. and Sumner, D. R. (1982) Prehelminthosporol and prehelminthosporol acetate: plant growth regulating properties. *J. Agric. Food. Chem.* **80**, 658-662.

Jennings, D. H. (1989) Some perspectives on nitrogen and phosphorus metabolism in fungi, in L. Boddy, R. Marchant and D. J. Read (eds.), *Nitrogen, Phosphorus and Sulfur Utilization by Fungi.* , Cambridge University Press, Cambridge, pp. 1-31.

Jørgensen, J. (1974) Occurence and importance of seed borne inoculum of *Cochliobolus sativus* on barley in Denmark. *Acta Agric. Scand.* **24**, 49-54.

Kurppa, A. (1984) *Bipolaris sorokiniana* on barley seed in Finland. *J. Agric. Sci. Finl.* **56**, 175-182.

Liljeroth, E., Franzon-Almgren, I. and Gustafsson, M. (1994) Effect of prehelminthosporol, a phytotoxin produced by *Bipolaris sorokiniana*, on barley root. *Can. J. Bot.* **72**, 558-563.

Markham, P. (1994) Occlusions of septal pores in filamentous fungi. *Mycol. Res.* **98**, 1089-1106.

Mason, P. J. and Crosse, R. (1975) Crystalline inclusions in hyphae of the *glaucus* group of *Aspergilli*. *Trans. Br. Mycol. Soc.* **65**, 129-134.

Mc Keen, W. E. (1971) Woronin bodies in *Erysiphe graminis* D.C. *Can. J. Microbiol.* **17**, 1557-1560.

Nilsson, P., Åkesson, H., Jansson, H.-B. and Odham, G. (1993) Production and release of the phytotoxin prehelminthosporol by *Bipolaris sorokiniana* during growth. *FEMS Microbiol. Ecol.* **102**, 91-98.

Olbe, M., Sommarin, M. Gustafsson, M. and Lundborg, T. (1995) Effect of the pathogen toxin prehelminthosporol (*Bipolaris sorokiniana*) on activities in barley root plasma membrane vesicles. *Plant Phytopathol.* In Press.

Panopoulos, N. J., Walton, J. D. and Willis, D. K. (1984) Genetic and biochemical basis of virulence in plant pathogens, in D. P. S. Verma and T. Hohn (eds.), *Plant Gene Research: Basic Knowledge and Application: Genes Involved in Microbe-Plant Interactions*, Springer Verlag, New York, pp. 339-374..

Pena-Rodriguez, L. M., Armingeon, N. A. and Chilton, W. S. (1988) Toxins from weed pathogens. I. Phytotoxins from a *Bipolaris* pathogen of Johnson grass. *J. Nat. Prod.* **51**, 821-828.

Sivanesan, A. and Holliday, P. (1981) *Cochliobolus sativus*. CMI Descriptions of pathogenic fungi and bacteria. 71 (701).

Tyner, L. E. (1966) Associative effects of fungi on *Cochliobolus sativus*. *Phytopathology* **56**, 776-780.

Wildermuth, G. B. and MacNamara, R. B. (1987) Susceptibility of winter and summercrops to root and crown infection by *Bipolaris sorokiniana*. *Plant Pathol.* **36**, 481-491.

MONITORING THE ACTIVITY OF DIFFERENT TRICHODERMA ISOLATES BY THE ISOELECTRIC POINTS (pI) OF THEIR EXTRACELLULAR ENZYMES

THRANE, C.[1], TRONSMO, A.[2], and JENSEN, D. F.[1]
[1] Plant Pathology Section, Department of Plant Biology, The Royal Veterinary and Agricultural University, Thorvaldsensvej 40, DK-1871 Frederiksberg C, Copenhagen, Denmark.
[2] Department of Biotechnological Sciences, Agricultural University of Norway, N-1432, Ås, Norway.

1. Abstract

The antagonistic *Trichoderma harzianum* isolates T3a and P1 were cultivated individually and in combination with and without the presence of the plant pathogen *Pythium ultimum* in sphagnum peat moss.

The number of colony forming units (cfu) of the *Trichoderma* isolates was determined on selective medium. Differentiation between the two isolates was possible by the use of T3a which is a GUS transformant of T3. Whether alone or in the presence of the pathogen, there were twice as many cfu of T3a than of P1. When T3a and P1 were cultivated together approx. 90 % of the cfu were T3a.

Proteins were extracted from the cultivations. ß-1,3-glucanase, exo- and endo ß-1,4-cellulase, and exo-and endo chitinase activities of these crude enzyme extracts were quantified. The activity that was most dramatically induced by the presence of the pathogen was ß-1,3-glucanase originating from isolate T3a, indicating that this activity is important in the antagonistic interaction between T3a and *P. ultimum*. The enzyme extracts from the different treatments were subjected to isoelectric focusing electrophoresis (IEF) and subsequently the IEF-gels were overlaid with a gel for detection of ß-1,3-glucanase activity. It was possible to distinguish the ß-1,3-glucanase produced by T3a and P1 by a small difference in their isoelectric points. In the combined cultivation of P1 and T3a, the majority of the ß-1,3-glucanase enzyme activity that could be detected had the pI of the enzyme produced by T3a. This result is consistent with the mentioned majority of cfu belonging to T3a when cultivated together with P1.

This result indicates that it is possible to monitor the activity of one specific fungal isolate when cultivated together with other isolates by combining quantitative

D. F. Jensen et al. (eds.), Monitoring Antagonistic Fungi Deliberately Released into the Environment, 105–111.
© 1996 Kluwer Academic Publishers. Printed in the Netherlands.

enzyme activity measurements and the use of activity gels for detection of specific hydrolytic enzymes.

2. Introduction

For monitoring of antagonistic fungi there are many available classical methods and new techniques. Several of these methods are based on general metabolic activity, others on DNA-probes which do not show whether the gene is expressed or not, etc. It would be advantageous to monitor organisms by measuring expressed biological factors that are specific and relevant for the organism and the environment in which it acts.

There are several reports on the importance of hydrolytic enzymes produced by antagonistic fungi for their biocontrol effect on plant pathogens either by a direct action on fungal cell walls (chitinase, ß-1,3-glucanase, protease) or through their ability to colonize roots or plant debris (cellulase, hemicellulase) (e.g. Lynch, 1989).

The two *Trichoderma harzianum* isolates P1 and T3 have been identified as biocontrol agents against plant pathogens. P1 was originally isolated from wood chips and is a cold resistant isolate that is able to control storage diseases at 4^0C (Tronsmo, 1989). This isolate controls many diseases caused by fungi which have chitin in their cell walls, but so far it has not been tested for its efficacy against the oomycetes. From this isolate chitinolytic enzymes have been purified and these have shown to inhibit germination of spores and hyphal growth of several plant pathogens in *in vitro* tests (Lorito *et al.*, 1993). T3a is a GUS-transformant (Thrane *et al.*, 1995a; Green and Jensen, 1995) of the wild type T3 which originally was isolated from *Pythium* suppressive sphagnum peat (Wolffhechel, 1989). From this isolate purified ß-1,3-glucanase and ß-1,4-cellulase have proven to inhibit growth of germ tubes from germinating spores and germination of cysts, respectively, from the oomycete *Pythium* (Thrane *et al.*, 1996a and 1996b). We were interested in comparing the competitive ability between the two isolates when cultivated in peat supplemented with straw with or without the presence of the plant pathogen *Pythium ultimum*. These cultivation conditions were likely to favour T3a.

The objective of this study was to develop a method to enable us to distinguish between enzymes produced by different antagonistic fungal isolates. This would be valuable as indication of the activity of a certain isolate when cultivated together with other antagonistic or plant pathogenic fungi.

Previous results comparing the pI's of ß-1,3 glucanase and ß-1,4 cellulase on enzyme activity gels from the *T. harzianum* isolates T3 and P1, *T. virens* (*Gliocladium virens*) isolates G2 and 62-6, and *G. roseum* (IK 726) showed surprisingly species-dependent variation in pI-values among these five isolates (unpublished).

3. Methods

3.1. FUNGAL STRAINS AND CULTIVATION CONDITIONS.

Trichoderma harzianum isolates T3a (GUS transformant of isolate T3) (Thrane *et al.*, 1995a; Green and Jensen, 1995) and P1 Tronsmo, 1989), *T. virens* (*Gliocladium virens*) isolates G2 and 62-6, and *G. roseum* IK726.

In flasks, agar plugs of fresh mycelium were incubated in peat with 2.5 % straw for 14 days at RT in the dark. Three replicates were made of each treatment (see Table 1 and 2 for the different treatments).

3.2. DILUTION PLATING.

Cfu counts were done by dilution plating on *Trichoderma* selective medium (TSM) (Elad *et al.*, 1981). Subsequently, colonies were tested for GUS-activity to distinguish between P1 and T3a (GUS-transformant) according to Thrane et al. (1995a).

3.3. MEASUREMENTS OF ENZYME ACTIVITIES.

Crude enzymes were extracted from the peat inoculations with 50 mM Na-phosphate buffer (pH 6.7) and subsequently separated from the solids by centrifugation and filtration. Cellulase and ß-glucanase activities were determined with blue substrates (Loewe Biochemica) according to Thrane *et al.* (1996a). ß-glucosidase, cellobiohydrolase, N-acetyl glucosaminidase (NAGáse) and chitobiosidase activity were determined by the use of p-nitrophenyl substrates (SIGMA) according to Thrane *et al.* (1996a) and Tronsmo and Harman (1993). Endochitinase activity was determined by measuring the reduction in turbidity of a solution of colloidal chitin according to Tronsmo and Harman (1993).

Concentrated crude enzymes were subjected to isoelectric focusing (IEF) and native SDS electrophoresis using premade gels (Pharmacia Phast system). After focusing the IEF-gel was placed on a ß-glucanase activity gel (1 % agarose, 50 mM Na-phosphate buffer, ß-glucanase blue substrate (50 %)).

Results

When P1 was cultivated by itself (treatments 2 and 4) there were about half the number of cfu's of T3a (treatments 3 and 5) (Table 1). In the mixed populations (treatments 6 and 7) 86-89 % of the cfu's were T3a.

In general higher enzyme activities were detected in the cultivations where T3a was present (Table 2). Enzyme activities that seem to be induced by the presence of *Pythium* are, in the case of P1, N-acetyl-ß-D-glucosaminidase (NAG'ase) and for T3a ß-1,3-glucanase, NAG'ase, and chitobiosidase.

On the ß-1,3-glucanase activity gel (*Figure 1*) it is seen that the small difference in pI of enzyme from the two isolates is sufficient to distinguish them (treatments 2-5). Further, in the treatments 6 and 7, where P1 and T3a are mixed, the major activity band has the pI of T3a, whereas only a slight P1 activity is seen on the gel.

TABLE 1. Colony forming units (cfu) on *Trichoderma* selective substrate (TSM). Percentage T3a of total cfu (= GUS-active cfu).

Treatment/ Fungi	cfu (10^8)	% GUS active cfu (=T3a)
1. *P. ultimum*	0.0 E	0
2. P1	5.1 CD	0
3. T3a	11.3 A	100
4. P1 + *P. ultimum*	4.6 D	0
5. T3a + *P. ultimum*	11.4 A	100
6. P1/T3a	8.4 B	89
7. P1/T3a + *P. ultimum*	7.3 BC	86

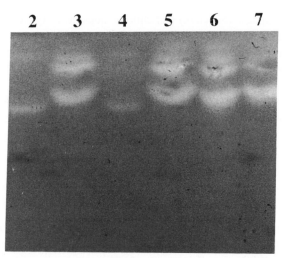

FIGURE 1. ß-glucanase activity gel after isoelectric focusing electrophoresis of extractable enzymes from treatments 2-7.   pI 4-6.5.

TABEL II. Activity measurements of extractable enzymes from the different treatments.

Treatment/ Fungi	protein (µg/ml)	cellulase (arb. units)	ß-glucanase (arb. units)	ß-glucosidase N katal	cellobiohydr. (N katal)	NAGáse (N katal)	chitobiosidase (N katal)	endochitinase (units)
1. *P.ultimum*	19.0 D	0.7 C	0.0 E	0.0 D	0.0 B	0.0 C	0.0 D	0.0 C
2. P1	59.7 C	8.5 B	5.5 D	62.7 C	0.0 B	57.3 C	106.3 C	0.7 B
3. T3a	108.3 A	18.8 A	8.0 B	156.7 AB	46.7 A	78.7 AB	149.3 BC	1.4 A
4. P1 + *P. ultimum*	60.0 C	8.9 B	4.8 D	67.3 C	0.0 B	76.7 AB	105.7 C	0.6 B
5. T3a + *P. ultimum*	118.0 A	20.5 A	22.3 A	179.7 A	54.3 A	88.7 A	212.3 A	1.6 A
6. P1/T3a	91.7 B	17.8 A	6.5 C	128.0 B	41.7 AB	81.0 AB	178.7 AB	0.5 B
7. P1/T3a + *P. ultimum*	87.0 B	18.2 A	6.7 BC	131.5 B	21.0 AB	78.3 AB	204.0 A	0.8 B

Discussion

Table I shows that T3a was a better competitor than P1 under the chosen cultivation conditions. This was expected as the wild type T3 initially was isolated from similar growth conditions, whereas P1 was isolated from wood. Some kind of competition (e.g. nutrients) could occur between the two *Trichoderma* isolates, as the numbers of cfu of both P1 and T3a were lower in the combined cultivations (approx. 15% P1 and 60% T3a). The presence of *Pythium* did not seem to cause any dramatic increase or decrease in the number of cfu.

The enzyme activity measurements shown in Table 2 are in agreement with the cfu determination (Table 1) indicating that T3a is more fit and active in the chosen environment. This is also reflected by the much higher activities of cellulose degrading enzymes when T3a is present.

In this experiment the most remarkable enzyme activity proved to be that of ß-glucanase. This enzyme is the only one that seems to be specifically induced in the T3a + *P. ultimum* cultivation (compare treatments 3 and 5 in Table 2). This result indicates that this enzyme is likely to be an antagonistic determinant for the T3a/*P. ultimum* antagonism (Thrane *et al.*, 1996b). Further, P1 has been selected for its high chitinolytic activity and biocontrol of plant pathogens with chitin containing cell walls, so it is possible that P1 has less pronounced effect on *P. ultimum* which does not have chitin as a structural component in the cell wall.

When comparing the quantity of ß-glucanase activity measured in the treatments 5 and 7 it could appear a little strange that the activity is much less in treatment 7 than in 5 (30%). However, this can be partly explained by the fact that there are less T3a cfu in treatment 7 than in treatment 5 (55%).

Due to the slight difference in the pI of the main ß-glucanase activity between P1 and T3a (*Figure 1*) we were able to determine which fungus that was responsible for the measured ß-glucanase activity in the different treatments. We were also able to exclude that *P. ultimum* was responsible for the detected ß-glucanase by making an activity gel after size separation of proteins by native gradient electrophoresis (comparison of activity bands in treatments 3, 5 & 7 in Table 2). In the treatments where P1 and T3a were mixed (6 & 7) it is obvious from figure 1. that the majority of the detected ß-glucanase activity originated from T3a. Further, on figure 1 a very weak P1a activity can be seen below the T3a main activity.

The results prove that it is possible to use this method to distinguish between activities of different isolates, when there are differences in the pI's of their extracellular enzymes. Further, in biocontrol systems it could be possible by this method to address the question of whether a specific enzyme activity is an antagonistic determinant.

References

Elad, Y., Chet, I., and Henis, Y. (1981) A selective medium for improving quantitative isolation of *Trichoderma* spp. from soil, *Phytoparasitica* **9**, 59-67.

Green, H. and Jensen, D. F. (1995) A tool for monitoring *Trichoderma harzianum*: II. GUS transformants used for ecological studies in the rhizosphere, *Phytopathology* 85, 1436-1440.

Lorito, M., Harman, G. E., Hayes, C. K., Broadway, R. M., Tronsmo, A., Woo, S. L. (1993) Chitinolytic enzymes produced by *Trichoderma harzianum*: Antifungal activity of purified endochitinase and chitobiosidase, *Phytopathology* **83**, 302-307.

Lynch, J. (1989) Environmental potential of the *Trichoderma* exocellular enzyme system, in N. G. Lewis and M. G. Paice (eds.), *Plant Cell Wall Polymers. Biogenesis and Biodegradation*, American Chemical Society, Washington D. C., pp. 608-618.

Thrane, C., Lübeck, M., Green, H., Degéfu, Y., Thrane, U., Allerup, S., and Jensen, D. F. (1995) A tool for monitoring *Trichoderma harzianum*: I Transformation with the GUS gene by protoplast technology, *Phytopathology* 85, 1428-1435.

Thrane, C., Tronsmo, A., and Jensen, D. F. (1996a *in preparation*) Identification, purification, and partial characterization of hydrolytic enzymes from an antagonistic isolate of *Trichoderma harzianum*.

Thrane, C., Tronsmo, A., and Jensen, D. F. (1996b *in preparation*) The role of a ß-1,3-glucanase and a ß-1,4-cellulase in the antagonistic interaction between *Trichoderma harzianum* isolate T3 and plant pathogenic *Pythium* spp.

Tronsmo, A. (1989) *Trichoderma harzianum* used for biological control of storage rot on carrots, *Norwegian J. Agric. Sci.* **3**, 157-161.

Tronsmo, A. and Harman, G. E. (1993) Detection and quantification of N-acetyl-ß-D-glucosaminidase, chitobiosidase, and endochitinase in solutions and on gels, *Anal. Biochem.* **208**, 74-79.

Wolffhechel, H. (1989) Fungal antagonists of *Pythium ultimum* isolated from a disease suppressive sphagnum peat, *Växtskyddsnotiser* **53**, 7-11.

PCR-BASED METHODS - A PROMISING TOOL FOR DETECTION AND IDENTIFICATION OF FUNGI IN SOIL.

METTE LUBECK AND PETER STEPHENSEN LUBECK

Department of Plant Biology, The Royal Veterinary and Agricultural University, 40 Thorvaldsensvej, DK-1871 Frederiksberg C, Denmark.

1. Introduction

The polymerase chain reaction (PCR) has the ability to amplify specific DNA sequences in an exponentially fashion by in vitro DNA synthesis (Mullis and Faloona, 1987; Saiki *et al.*, 1988). In principle, it is possible from just one or a few copies of target DNA to amplify sufficient amounts of DNA in a few hours to enable visualization of DNA by ethidium bromide staining in gel electrophoresis. Because of these features, PCR is now considered to be a very important technique in molecular biology and has found increasing applications in many different disciplines. In molecular ecology, the technique can be used for producing sensitive and simple methods of detecting and identifying organisms in the environment due to the potential of PCR to detect a few target sequences in a complex mixture. The specificity of the method is dependent on design of oligonucleotide primers, and as with serology, both narrow and broad selectivities are possible. This allows differentiation at many taxonomic levels, and monitoring of specific populations or isolates (Henson and French, 1993; Ward, 1994).

When developing a technique for detection and identification of organisms in a complex environment such as soil one of the most central aspects to be considered is the specificity of the technique. The target group might be at genus level, species level or isolate level. Regardless of the level, the essential in detection is to be able to detect all members of the target group (eliminating all false-negative) and discriminate all other organisms (eliminating all false-positive). This requirement of specificity can in most cases only be full-filled by knowledge of genetic relatedness of the target organism(s) with other organisms. Therefore the important preliminary step in developing a PCR assay for organism(s) of interest is to study their genetic diversity to be able to form reliable groups as well as their genetic affinity with all relevant organisms.

With fungi, one of the first uses of PCR was to amplify ribosomal DNA sequences for determination of phylogenetic relationships (White *et al.*, 1990), but now PCR is used in many other areas of mycology including detection and identification of fungi (for a review, see Foster *et al.*, 1993). This mini-review describes some of the most

D. F. Jensen et al. (eds.), Monitoring Antagonistic Fungi Deliberately Released into the Environment, 113–121.

applicable approaches of PCR techniques in detection and identification of fungi: 1) analysis of ribosomal RNA genes (rDNA), 2) specific detection by PCR, and 3) fingerprint techniques such as random amplified polymorphic DNA (RAPD) and universally primed PCR (UP-PCR).

2. Analysis of Ribosomal DNA.

Nuclear ribosomal RNA genes (rDNA) are tandemly repeated multigene families containing both coding and noncoding (spacer) regions. Each repeat unit in eukaryotes contains a copy of the 18S, 5.8S, and 28S rDNA, and two internal transcribed spacers, ITS1 and ITS2, flanking the 5.8S gene (White *et al.*, 1990). The regions of the 18S, the 25S, and the 5.8S are much more conserved in nucleotide sequences compared to the ITS1 and ITS2 regions. The rDNA regions are attractive as targets for PCR because they are present in up to 220 copies in the fungal genome, and they can be used to differentiate between genera, species and subspecies depending on which rDNA regions that are targets for investigations (Bruns *et al.*, 1991).

The presence of phylogenetically conserved regions in the rDNA, have facilitated the development of primers (hereafter called conserved primers) which are useful in most fungi (White *et al.*, 1990). Some of the conserved primer sites flank variable regions allowing specific amplification of these. Because of the features of rDNA, these regions represent an attractive target for many different investigations such as analysis of phylogenetic relationships (White *et al.*, 1990; Bruns *et al.*, 1991), and for identification and detection of fungi using different approaches.

2.1. RESTRICTION ANALYSIS OF AMPLIFIED RIBOSOMAL DNA.

By using conserved primers it is possible to selectively amplify regions of ribosomal DNA and obtain fragments of similar size in most fungi. Since rDNA contains areas with some nucleotide variability, the amplified regions are targets for digestion with restriction enzymes followed by analysis of restriction fragment length polymorphism (RFLP) in gel electrophoresis. RFLP pattern obtained by this strategy can be used as fingerprints. Because of the high degree of conservation of rDNA and the amount of restriction enzymes available, these fingerprints is in most cases limited to species rather than isolates compared with the fingerprinting methods RAPD and UP-PCR, described in section 4, which have stronger potential for discrimination at isolate level. However, because the ITS1 and ITS2 regions are more variable in nucleotide sequence composition than either the 18S, 5.8S or 25S, and may also vary in number of nucleotides due to insertions and deletions (Bryan *et al.*, 1995), it can be possible to distinguish groups of isolates with different ITS size without any need for sequencing (see for example Lübeck *et al.*, 1996). Many have found, that restriction analysis involving the ITS regions allow discrimination of fungi at the species (Chen *et al.*, 1992; Brown *et al.*, 1994; Ward *et al.*, 1994) and sometimes subspecies level (Liu and Sinclair, 1992).

The RFLP pattern for a range of organisms closely related can furthermore be used as a base for determination of relationships between the organisms (Holsinger and Jansen, 1993). Restriction enzyme analysis that involves at least in part the conserved regions are most suitable for inferring of phylogenetic relationships (Bruns *et al.*, 1991; Bulat, *pers. comm.*).

2.2. SEQUENCING ANALYSIS OF RIBOSOMAL DNA.

The high variability in the ITS regions make these attractive targets for inferring phylogenies (Bruns et al., 1991) and PCR detection of target organism(s) through primer development (Mills *et al.*, 1992; Brown *et al.*, 1993; Tisserat *et al.*, 1994). Also other parts of the ribosomal DNA region has been used for primer development, e.g. the 18S region (Simon *et al.*, 1992), mitochondrial small subunit rDNA (Li *et al.*, 1994), and the intergenic spacer (IGS) regions (Holmes *et al.*, 1994).

By amplifying and direct sequencing the corresponding regions from different but related organisms, it is possible from alignment of the sequences to design specific primers directed to areas with non-identical sequences (Doolittle, 1990; Wheeler, 1994). Development of such primers can be used in specific detection by PCR, see section 3.

Sequences of ribosomal DNA can be analyzed for relatedness between the sequences by specially designed computer programs (Hillis *et al.*, 1993). This information is useful for determination of phylogenetic relationships of the organisms (White *et al.*, 1990; Bruns *et al.*, 1991; Morton *et al.*, 1995). Sequences from the ITS regions can also be used directly for identification of isolates that are indistinguishable by traditional morphological and physiological characters (Boysen *et al.*, 1996; Rossen, *pers. comm.*). However, as differences in ITS sequence can be very small between well characterized species, sequence data are more powerful when used in combination with other methods for identification (Boysen *et al.*, 1996).

3. Specific Detection by PCR.

Specific detection by PCR differs from the rDNA amplifications (section 2) and the PCR fingerprinting techniques (section 4) in that it requires knowledge of the target DNA sequences for development of specific primers. Most often two specific primers are used to amplify one fragment of known size (a diagnostic product) recognized by gel electrophoresis. To develop such primers it is necessary to identify DNA sequences that are unique to the organism. These sequences can be found by different approaches, for example by using specific probes that already exist for detection of the organism (Lanfranco *et al.*, 1993; Schaad *et al.*, 1995), by subtractive hybridization (Seal *et al.*, 1992), by knowledge of specific genes with unique sequences (Burgener-Kairuz *et al.*, 1994; Lamar *et al.*, 1995), by using information obtained by sequencing the variable ITS region (section 2.2), or by sequencing unique amplified fragments from the fingerprint methods (section 4).

Design of primers from a specific gene require knowledge of sequences of the gene in the target organism(s) and knowledge of sequences of corresponding genes in related organisms. In some cases published sequences can be used solely as the basis to design primers from highly variable regions, but in many cases only few sequences are available from such a target gene. One strategy which can be used, is to align known sequences (maybe only two or three) (Doolittle, 1990; Wheeler, 1994), design primers (amplimers) from the conserved regions and use these to amplify the corresponding region of related organisms including the target organism. By direct sequencing and subsequent alignment of the sequences, it may be possible from variable regions to design specific primers for the target organism. Use of these regions for primer development depends on the existence of adequate sequence difference to design primers which specifically amplify DNA from the target organism(s). Primers based on single base differences will probably not be robust since single mutations could undermine the differentiation (Morton *et al.*, 1995). After design of specific primers, it is necessary to check the specificity i.e. that they allow amplification of only the desired organism and not of other related or unrelated fungi (section 5).

4. Random Amplified Polymorphic DNA (RAPD) And The Universally Primed PCR (UP-PCR) Technique.

The random amplified polymorphic DNA (RAPD) (Welsh and McClelland, 1990; Williams *et al.*, 1990) and the universally primed PCR (UP-PCR) technique (Bulat and Mironenko, 1990; Bulat *et al.*, 1994) are similar in that it is possible to amplify DNA from any organism without previous knowledge of DNA sequences. With both methods, the use of one single primer provides amplification of several DNA fragments randomly distributed throughout the genome.

The UP-PCR technique differs from the more well-known RAPD technique in the design of primers. The primers used in RAPD are usually short, 8-12-mers, with random sequence composition, and the amplification process is very sensitive to reaction conditions, in particular the annealing temperature. In contrast, the primers used in UP-PCR (UP primers) usually are longer, 13-21-mer, and have a unique design. They contain two parts, of which the 3' end (up to 8 nucleotides) consists of random sequences that do not match sequences of phylogenetic conserved loci (like rDNA), and the 5' end, is universal, meaning that used alone this allows amplification of DNA from any organism. Based on the UP-PCR technique, different derivative methods have been developed (Bulat, *pers. comm.*). One of these, the Species Identification method, is a cross-blot hybridization variant of UP-PCR, where hybridization of UP-PCR products obtained from isolates of one species with the same primer reveals DNA homology (Bulat and Mironenko, 1992; Bulat *et al.*, 1994; Lübeck *et al.*, 1996). Other variants are the Phylogenetic Relationships Inference method, and development of specific diagnostics by using pairwise combinations of selected primers for amplification of a specific product (Bulat, *pers. comm.*).

Where an amplification product is found to be unique for a target organism/group, the RAPD and UP-PCR techniques can be used for design of probes or specific primers (Lanfranco *et al.*, 1993; Ménard *et al.*, 1994; Bulat, *pers. comm.*).

RAPD and UP-PCR analysis using one primer is not suitable for direct detection of target organisms in complex environmental samples. Both methods require pure cultures because of the nature of the primers that allows amplification from almost any organisms. The main advantage of the methods is that both of them have potential to differentiate very similar isolates of the same species by fingerprinting which make the methods useful in population studies and in recognition of isolates (see for example Lübeck *et al.*, 1996). However, a variant of the UP-PCR technique with selected combinations of UP primers which can amplify a specific diagnostic product in the presence of foreign DNA open the possibility for direct detection of organisms in complex environments as soil (Bulat, *pers. comm.*).

5. Discussion.

Many factors are necessary to take into account when choosing a method for detecting or identifying specific fungal isolates in soil. It should be remembered that for all methods the sampling procedure chosen is crucial and is often a limiting factor in the detection of specific microorganisms due to their behavior in soil (Tsushima *et al.*, 1995). Also important to have in mind is the possible outcome of the different methods. Methods like RAPD, UP-PCR, and rDNA restriction analysis require in most cases template DNA from pure cultures (homogeneous DNA from one isolate). Detection or identification based on such methods require then standard methods like plate-dilution techniques prior to amplification. Sensitivity of such methods are thus equivalent to/or limited to the sensitivity of the plate-dilution method. In contrast, detection by PCR with specific primers can be used more directly in complex environmental samples (heterogeneous DNA from mixtures of isolates), and the sensitivity is then related to other parameters. However, as direct detection do not discriminate between live or dead material, it may be necessary to combine PCR with other strategies to evaluate the presence of viable fungal material. Some strategies to solve this may be using methods like BIO-PCR (combined biological and enzymatic amplification) where only living cells are detected (Schaad *et al.*, 1995), or reverse transcriptase PCR (RT-PCR) where the target for detection are specific mRNAs allowing detection of activity of the organisms (Lamar *et al.*, 1995). In contrast to BIO-PCR, RT-PCR does not detect resting spores and structures.

Detection of small numbers of target organisms requires high level of sensitivity. There are many different ways to optimize PCR for detection of low numbers of cells in environmental samples. These include number of target gene copies, concentration of cells from samples, complexity of abiotic and biotic components, and extraction and purification of DNA from sediment and soil, the latter often being the most difficult step. Many have found that additional cycles of amplification enhance sensitivity (Steffan and Atlas, 1991; Bej and Mahbubani, 1992) but according to Henson and French (1993), Bulat (*pers. comm.*) and own experiences too many cycles increases the risk of contaminating reactions (accumulation of non-specific PCR-amplified products).

Another strategy to increase sensitivity is by using labelled primers or probes to detect non-visible amplification products (Steffan and Atlas, 1988; Schraft and Griffiths, 1995). Many authors have found that the enhanced sensitivity obtained by hybridization can be replaced by the use of nested primer PCR which is faster and more simple (Barlough *et al.*, 1994; Catalan *et al.*, 1994; Straub *et al.*, 1994). Use of nested primers involves a second amplification reaction with different primers in the second reaction which are internal to the first set (Mullis and Faloona, 1987).

A main problem with PCR detection is the possibility of obtaining false negative, i.e. positive samples that responds negatively in the PCR reaction. This can be due to a number of factors. One way of checking that the DNA is of suitable quality and quantity for PCR is to test the DNA in a separate experiment using conserved primers (Ward, 1994). Another strategy could be to include an internal control in the reaction (duplex PCR), for example by adding an extra set of primers which are able to amplify most fungi (Ward, 1995).

A single pair of primers does not in all cases possess sufficient specificity, where other non-related organisms might also be detected (false-positive). There are several ways of overcoming the risk of obtaining such false positive, one of these are by using nested PCR methods (McManus and Jones, 1995; Tsushima *et al.*, 1995), another could be to increase the annealing temperature to ensure that the primers only bind to homologous primer sites (Ward, 1994; Ward, 1995).

Most PCR protocols in detection are qualitative, i.e. have the ability to detect the presence of the target organisms but do not quantitate the target organisms. There have been many efforts to use PCR for quantification (see for example Hu *et al.*, 1993; Degrange and Bardin, 1995; Lamar *et al.*, 1995) but great care in optimization of reaction conditions and interpretation of the results have to be taken. Minor differences in the efficiency of the reaction between samples can give rise to tremendously different amounts of final product, since PCR amplification proceeds exponentially. One of the frequently used quantitative methods is inclusion of known amounts of internal control sequences (competitors) in the PCR reactions (Gilliland *et al.*, 1990; Lamar *et al.*, 1995). Using this strategy it is possible to reproducibly quantify the amount of target DNA in an environmental sample by titrating unknown amounts of target DNA against dilution series of competitor DNA (Gilliland *et al.*, 1990; Steffan and Atlas, 1991).

In conclusion, PCR has potential to be used in many different ways to identify and detect fungi in soil. At present, one of the main problems are the risk of obtaining false-positive or false-negative because the PCR reaction is very sensitive to small changes in the involved parameters. Determination of genetic relationship of the target organism(s) with other related organisms is important information for evaluation of the versatility of the method and is necessary when evaluating the specificity of developed primers for use in specific detection.

Acknowledgements.

We are grateful to Dr. Sergey Bulat and Dr. Lone Rossen for critically reading and commenting the manuscript.

References

Barlough, J., East, N., Rowe, J.D., Hoosear, K.V., DeRock, E., Bigornia, L., and Rimstad, E. (1994) Double-nested polymerase chain reaction for detection of caprine arthritis-encephalitis virus proviral DNA in blood, milk, and tissues of infected goats, *Journal of Virological Methods* 50, 101-114.

Bej, A.K., and Mahbubani, M.H. (1992) Applications of the polymerase chain reaction in environmental microbiology, *PCR Methods and Applications* 1, 151-159.

Boysen, M., Skouboe, P., Frisvad, J., and Rossen, L. (1996) Reclassification of the *Penicillium roqueforti* group into three species on the basis of molecular genetic and biochemical profiles, *Microbiology, in press*.

Brown, A.E., Muhthumeenakshi, S., Sreenivasaprasad, S., Mills, P.R., and Swinburne, T.R. (1993) A PCR primer-specific to *Cylindrocarpon heteronema* for detection of the pathogen in apple wood, *FEMS Microbiology Letters* 108, 117-120.

Brown, A.E., Muthumeenakshi, S., Swinburne, T.R., and Li, R. (1994) Detection of the source of infection of apple trees by *Cylindrocarpon heteronema* using DNA polymorphisms, *Plant Pathology* 43, 338-342.

Bruns, T.D., White, T.J., and Taylor, J.W. (1991) Fungal molecular systematics, *Annual Review of Ecology and Systematics* 22, 525-564.

Bryan, G.T., Daniels, M.J., and Osbourn, A.E. (1995) Comparison of fungi within the *Gaeumannomyces-Phialophora* complex by analysis of ribosomal DNA sequences, *Applied and Environmental Microbiology* 61, 681-689.

Bulat, S.A., and Mironenko, N.V. (1990) Species identity of the phytopathogenic fungi *Pyrenophora teres* Drechsler and *P. graminea* Ito and Kuribayashi, *Mikologia i Phytopathologia* (Russ.) 24, 435-441.

Bulat, S.A., and Mironenko, N.V. (1992) Polymorphism of yeast-like fungus *Aureobasidium pullulans* (De Bary) revealed by universally primed polymerase chain reaction: species divergence state, *Genetika* (Russ.) 28, 19-30.

Bulat, S.A., Mironenko, N.V., Lapteva, M.N., and Strelchenko, P.P. (1994) Polymerase chain reaction with universal primers (UP-PCR) and its application to plant genome analysis, in R.P. Adams *et al.* (eds.), *Conservation of Plant Genes II: Utilization of ancient and modern DNA*, Missouri Botanical Garden, St. Louis 48, 113-129.

Burgener-Kairuz, P., Zuber, J.P., Jaunin, P., Buchman, T.G., Bille, J., and Rossier, M. (1994) Rapid detection and identification of *Candida albicans* and *Torulopsis (Candida) glabrata* in clinical specimens by species-specific nested PCR amplification of a cytochrome P-450 lanosterol-α-demethylase (L1A1) gene fragment, *Journal of Clinical Microbiology* 32, 1902-1907.

Catalan, V., Moreno, C., Dasi, M.A., Munoz, C. and Apraiz, D. (1994) Nested polymerase chain reaction for detection of *Legionella pneumophila* in water, *Research in Microbiology* 145, 603-610.

Chen, W., Hoy, J.W., and Schneider, R.W. (1992) Species-specific polymorphism in transcribed ribosomal DNA of five *Pythium* species, *Experimental Mycology* 16, 22-34.

Degrange, V., and Bardin, R. (1995) Detection and counting of *Nitrobacter* populations in soil by PCR, *Applied and Environmental Microbiology* 61, 2093-2098.

Doolittle, R.F. (1990) Molecular evolution: computer analysis of protein and nucleic acid sequences, *Methods in Enzymology* 183

Foster, L.M., Kozak, K.R., Loftus, M.G., Stevens, J.J., and Ross, I.K. (1993) The polymerase chain reaction and its application to filamentous fungi, *Mycological Research* 97, 769-781.

Gilliland, G., Perrin, S., and Bunn, H. (1990) Competitive PCR for quantitation of mRNA, in M.A. Innis, D.H. Gelfand, J.J. Sninsky, and T.J. White (eds.) *PCR Protocols. A Guide to Methods and Applications*, Academic Press, San Diego, 60-69.

Henson, J.M., and French, R. (1993) The polymerase chain reaction and plant disease diagnosis, *Annual Review of Phytopathology* 31, 81-109.

Hillis, D.M., Allard, M.W., and Miyamoto, M.M. (1993) Analysis of DNA sequence data: phylogenetic inference, in E.A. Zimmer, T.J. White, R.L. Cann, and A.C. Wilson (eds.), *Molecular Evolution: Producing the Biochemical Data, Methods in Enzymology* 224, 456-487.

Holmes, A.R., Cannon, R.D., Shepherd, M.G., and Jenkinson, H.F. (1994) Detection of Candida albicans and other yeasts in blood by PCR, *Journal of Clinical Microbiology* 32, 228-231.

120

Holsinger, K.E., and Jansen, R.K. (1993) Phylogenetic analysis of restriction site data, in E.A. Zimmer, T.J. White, R.L. Cann, and A.C. Wilson (eds.), *Molecular Evolution: Producing the Biochemical Data, Methods in Enzymology* **224**, 439-455.

Hu, X., Nazar, R.N., and Robb, J. (1993) Quantification of *Verticillium* biomass in wilt disease development, *Physiological and Molecular Plant Pathology* **42**, 23-36.

Lamar, R.T., Schoenike, B., Vanden Wymelenberg, A., Stewart, P., Dietrich, D.M., and Cullen, D. (1995) Quantitation of fungal mRNAs in complex substrates by reverse transcription PCR and its application to *Phanerochaete chrysosporium*-colonized soil, *Applied and Environmental Microbiology* **61**, 2122-2126.

Lanfranco, L., Wyss, P., Marzachi, C., and Bonfante, P. (1993) DNA probes for identification of the ectomycorrhizal fungus *Tuber magnatum* Pico, *FEMS Microbiology Letters* **114**, 245-252.

Li, K.-N., Rouse, D.I., and German, T.L. (1994) PCR primers that allow intergeneric differentiation of ascomycetes and their application to *Verticillium* spp., *Applied and Environmental Microbiology* **60**, 4324-4331.

Liu, Z.L., and Sinclair, J.B. (1992) Genetic diversity of *Rhizoctonia solani* anastomosis Group 2, *Phytopathology* **82**, 778-787.

Lübeck, M., Bulat, S.A., Lübeck, P.S., Mironenko, N., and Jensen, D.F. (1996) Identification and characterization of isolates of *Trichoderma* and *Gliocladium* by PCR-based methods, *this volume*.

Ménard, C., Gosselin, P., Duhaime, J.-F., and Mouton, C. (1994) Polymerase chain reaction using arbitrary primer for the design and construction of a DNA probe specific for *Porphyromonas gingivalis*, *Research in Microbiology* **145**, 595-602.

Mills, P.R., Sreenivasaprasad, S., and Brown, A.E. (1992) Detection and differentiation of *Colletotrichum gloeosporioides* isolates using PCR, *FEMS Microbiology Letters* **98**, 137-144.

Morton, A., Garder, J.H., and Barbara, D.J. (1995) Sequences of the internal transcribed spacers of the ribosomal RNA genes and relationships between isolates of *Verticillium alboatrum* and *Verticillium dahliae*, *Plant Pathology* **44**, 183-190.

Mullis, K.B., and Faloona, F.A. (1987) Specific synthesis of DNA in vitro via a polymerase-catalyzed chain reaction, *Methods in Enzymology* **155**, 335-351.

Saiki, R.K., Gelfand, D.H., Stoffel, S., Scharf, S.J., Higuchi, R., Horn, G.T., Mullis, K.B., and Erlich, H.A. (1988) Primer-directed enzymatic amplification of DNA with a thermostable DNA polymerase, *Science* **239**, 487-491.

Schaad, N.W., Cheong, S.S., Tamaki, S., Hatziloukas, and Panopoulos, N.J. (1995) A combined biological and enzymatic amplification (BIO-PCR) technique to detect *Pseudomonas syringae* pv. *phaseolicola* in bean seed extracts, *Phytopathology* **85**, 243-248.

Schraft, H., and Griffiths, M.W. (1995) Specific oligonucleotide primers for detection of lecithinase-positive *Bacillus* spp. by PCR, *Applied and Environmental Microbiology* **61**, 98-102.

Simon, L., Lalonde, M., and Bruns, T.D. (1992) Specific amplification of 18S fungal ribosomal genes from vesicular-arbuscular endomycorrhizal fungi colonizing roots, *Applied and Environmental Microbiology* **58**, 291-295.

Steffan, R.J., and Atlas, R.M. (1988) DNA amplification to enhance detection of genetically engineered bacteria in environmental samples, *Applied and Environmental Microbiology* **54**, 2185-2191.

Steffan, R.J., and Atlas, R.M. (1991) Polymerase chain reaction: applications in environmental microbiology, *Annual Review of Microbiology* **45**, 137-162.

Straub, T.M., Pepper, I.L., Abbaszadegan, M., and Gerba, C.P. (1994) A method to detect enteroviruses in sewage sludge-amended soil using the PCR, *Applied and Environmental Microbiology* **60**, 1014-1017.

Tisserat, N.A., Hulbert, S.H., and Sauer, K.M. (1994) Selective amplification of rDNA internal transcribed spacer regions to detect *Ophiosphaerella korrae* and *O. herpotricha*, *Phytopathology* **84**, 478-482.

Tsushima, S., Hasebe, A., Komoto, Y., Carter, J.P., Miyashita, K., Yokoyama, K., and Pickup, R.W. (1995) Detection of genetically engineered microorganisms in paddy soil using a simple and rapid "nested" polymerase chain reaction method, *Soil Biology and Biochemistry* **27**, 219-227.

Ward, E. (1994) Use of the polymerase chain reaction for identifying plant pathogens, in J.P. Blakeman and B. Williamsen (eds.), *Ecology of Plant Pathogens*, CAB International, 143-160.

Ward, E. (1995) Improved polymerase chain reaction (PCR) detection of *Gaumannomyces graminis* including a safeguard against false negatives, *European Journal of Plant Pathology* **101**, 561-566.

Welsh, J., and McClelland, M. (1990) Fingerprinting genomes using PCR with arbitrary primers, *Nucleic Acids Research* **18**, 7213-7218.

Wheeler, W.C. (1994) Sources of ambiguity in nucleic acid sequence alignment, in B. Schierwater, B. Streit, G.P. Wagner, and R. DeSalle (eds.) *Molecular Ecology and Evolution: Approaches and Applications*, Birkhäuser Verlag, Basel, 323-352.

White, T.J., Bruns, T.D., Lee, S., and Taylor, J.W. (1990) Amplification and direct sequencing of fungal ribosomal RNA genes for phylogenetics, in M. Innis, D.H. Gelfand, J.J. Sninsky, and T.J. White (eds.), *PCR Protocols: a Guide to Methods and Applications*, Academic Press, San Diego, pp. 315-322.

Williams, J.G.K., Kubelik, A.R., Livak, K.J., Rafalski, J.A., and Tingey, S.V. (1990) DNA polymorphisms amplified by arbitrary primers are useful as genetic markers, *Nucleic Acids Research* **18**, 6531-6535.

IDENTIFICATION AND CHARACTERIZATION OF ISOLATES OF *TRICHODERMA* AND *GLIOCLADIUM* BY PCR-BASED METHODS.

METTE LUBECK[1], SERGEY A. BULAT[2], PETER STEPHENSEN LUBECK[1], NINA MIRONENKO[3], AND DAN FUNCK JENSEN[1].
1. *Department of Plant Biology, The Royal Veterinary and Agricultural University, Thorvaldsensvej 40, DK-1871 Frederiksberg C, Denmark.*
2. *Petersburg Nuclear Physics Institute, Gatchina, 188350, Russia.*
3. *All-Russian Plant Protection Institute, Podbelsky str. 3, Saint-Petersburg, 189620, Russia.*

Summary

We have analyzed 41 isolates belonging to species of *Trichoderma* and *Gliocladium* fungi, respectively, using two different techniques, Universal Primed PCR (UP-PCR) and mobility of the internal transcribed spacer, ITS1, of ribosomal DNA. The use of one universal primer (L45) gave a specific polymorphism in gel electrophoresis for most isolates and facilitated their differentiation. Dot-blot hybridization where UP-PCR products from one isolate hybridized to UP-PCR products of different isolates, distinguished the isolates as genomic species. The ITS1 regions from all isolates were amplified by using two specially designed primers and analyzed in gel electrophoresis. The difference in their mobility was used for isolate grouping. The UP-PCR and ITS1 groups revealed, proved to be consistent.

1. Introduction

Species from the fungal genera *Trichoderma* and *Gliocladium* are known to be antagonistic to other fungi, and have shown promise as biological control agents of several soilborne diseases (Jensen and Wolffhechel, 1995; Papavizas, 1985). The taxonomy of these genera is not fully clarified (Bissett, 1991; Domsch *et al.*, 1980), and there is great confusion about correct species designation. For example, one species has been identified as *G. virens*, partly through its ability to produce gliotoxin and viridin (Webster and Lomas, 1964). Ribosomal DNA data suggests, however, that the species should be placed within *Trichoderma* rather than *Gliocladium* (Rehner and Samuels, 1994).

The genus of *Trichoderma* has been classified into species aggregates based on

123

D. F. Jensen et al. (eds.), Monitoring Antagonistic Fungi Deliberately Released into the Environment, 123–128.
© 1996 *Kluwer Academic Publishers. Printed in the Netherlands.*

morphological characteristics (Rifai, 1969), but the classification suffers from large variation both within and between each aggregate. A number of biological species may be present in each aggregate, and morphological or genetic characters that reliably define these species within the aggregates have not been identified (Bissett, 1984). Bissett (1991) proposed sections based on morphology to accommodate similar forms within the species aggregates of Rifai (1969), but the morphological differences can be very difficult to recognize. Obviously, there is a urgent need for a more robust system to differentiate among these phenotypically similar species.

Many different methods have been employed for classification: restriction fragment length polymorphism (RFLP) (Meyer *et al.*, 1992), random amplified polymorphic DNA (RAPD) (Muthumeenakshi *et al.* 1994; Zimand *et al.*, 1994), sequencing of the internal transcribed spacer (ITS1) of rDNA (Muthumeenakshi *et al.* 1994), and analysis of mitochondrial DNA (Meyer, 1991), of secondary metabolites (Ghisalberti and Sivasithamparam, 1991), and of isozymes (Stasz *et al.*, 1989). Some of these methods have resulted in division of aggregates into subgroups, for example *T. harzianum* was divided into 3 groups by RAPD pattern and ITS1 sequences (Muthumeenakshi *et al.* 1994), and *T. viride* was divided into two groups on the basis of restriction enzyme profiles of mitochondrial DNA, and different conidial ornamentation revealed by scanning electron microscopy (Meyer, 1991; Meyer and Plaskowitz, 1989).

Dilution plating on selective media have traditionally been used for population studies of isolates of *Trichoderma* and *Gliocladium* (Papavizas, 1985). These methods do not distinguish isolates of the same species, and often not among species groups. Therefore, identification methods for specific isolates are necessary. A method which can be used for distinguishing different isolates will allow us to monitor and study the behavior of introduced isolates. This is important when evaluating survival and establishment of the isolates, and with respect for risk evaluation in relation to deliberately release of isolates for commercial purposes (Jensen and Wolffhechel, 1995). Such a method will also facilitate studies of the effect of selection pressure on diversity of indigenous *Trichoderma* and *Gliocladium* isolates in soil and rhizosphere. This could provide us with more knowledge of the genetic structure, and population dynamics of indigenous fungal populations in different soil types before introducing selected isolates. Information about content of various isolates in suppressive versus non-suppressive soils could assist in clarifying whether some isolates are related to the suppressiveness, and, thus, could be used as bio-indicators.

The objectives of our research are identification and characterization of isolates of *Trichoderma* and *Gliocladium* with the UP-PCR method and ribotyping, to determine their phylogenetic relationships, to develop specific molecular markers, and a PCR diagnostic method for antagonistic fungal isolates. This could be useful for clarifying species status of the isolates, for studying diversity among isolates of *Trichoderma* and *Gliocladium* as well as for developing a PCR based assay to monitor selected isolates in soil and rhizosphere.

For these reasons, we at first intended to choose a technique which enable the separation of different isolates facilitating reliable identification. At present we have analyzed 41 isolates belonging to 7 species of *Trichoderma* and *Gliocladium* fungi using two different techniques: Universal Primed PCR (UP-PCR), and mobility shift analysis of ITS1 ribosomal DNA (rDNA) (Bulat *et al.*, *unpublished data*).

2. Universal Primed PCR (UP-PCR)

The UP-PCR technique is similar to the well-known RAPD technique (Williams et al., 1990) in that it is possible to amplify DNA from any organism without previous knowledge of DNA sequences (Bulat *et al.*, 1994; Lübeck and Lübeck, 1996). The result of the reaction is amplification of several DNA fragments randomly distributed throughout the genome. The fragments are visualized by ordinary acrylamide gel electrophoresis, and provides direct analysis of polymorphism of different isolates. The UP-PCR technique can give useful information on variation both within and between species.

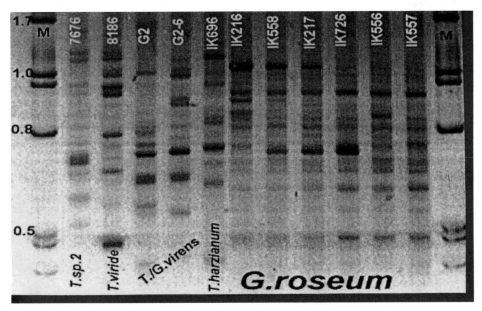

Figure 1. UP-PCR patterns of isolates of Trichoderma and Gliocladium with primer L45 (lanes 2-12). Markers M are lambda digested with Pst1 (lane 1 and 13).

As an example, we show the polymorphism obtained with one selected universal primer L45 (Figure 1) which gave a specific banding profile in gel electrophoresis for almost all isolates and allowed us to distinguish them from each other (Bulat *et al.*, *unpublished data*). Use of the UP-PCR polymorphisms alone for species designation of each isolate proved insufficient in many cases (except for *G. roseum*) because of high variation in UP-PCR banding pattern of isolates. Therefore, we used a hybridization variant of the UP-PCR technique (Bulat *et al.*, 1992) where all amplification products from one isolate were labelled and hybridized to UP-PCR products from the different isolates in cross dot-blot hybridization. The strength of the hybridization signal indicated the relatedness of the isolates to the tested isolate (used as a labelled probe). From these results, the isolates under study were grouped into genomic species (Bulat *et al.*, *unpublished data*).

3. Mobility Shift Analysis of ITS1 Ribosomal DNA

Ribosomal DNA genes are tandemly repeated multigene families containing both genic and nongenic (spacer) regions. Each repeat unit contains a copy of the 18S, 5.8S, and 28S rDNA, and two internal transcribed spacers, ITS1 and ITS2, flanking the 5.8S gene. Comparative studies of the nucleotide sequences of ribosomal RNA genes provide a means for analyzing phylogenetic relationships over a wide range of taxonomic levels (Bruns *et al.*, 1991; White *et al.*, 1990). Some of the regions of the rRNA genes are very conserved, e.g. most of the nuclear small-subunit rDNA (18S), the nuclear large-subunit (25S), and the 5.8S subunit, while others are more variable, e.g. the mitochondrial rRNA genes, and the internal transcribed spacers ITS1 and ITS2.

In this preliminary study, we describe a mobility shift assay for the ITS1 sequence of the ribosomal DNA. To amplify this region, two primers of our design were used, one (Y) prime DNA at the 3' region of the nuclear 18S subunit rDNA, and the other (X) prime at the 5.8S subunit. The mobility (size) of the ITS1 region were analyzed directly, and after digestion with one restriction enzyme (*Sau*3A) by acrylamide gel electrophoresis. The data obtained was processed by using original software for scanning analysis (Petersburg Nuclear Physics Institute, Russia) to obtain a more precise determination of the differences in ITS1 spacer size, and in the pattern of the *Sau*3A digestions (Bulat *et al.*, *unpublished data*). Based on differences in ITS1 mobility and *Sau*3A digestion patterns, it was possible to group the isolates in seven ITS1 groups. The ITS1 groups generally proved to be consistent with the genomic species obtained by the UP-PCR method (Bulat *et al.*, *unpublished data*).

4. Conclusion and Future Plans

Our preliminary results obtained for 41 isolates from seven species of *Trichoderma* and *Gliocladium* with UP-PCR and analysis of the ITS1 rDNA region indicates the usefulness of these methods for our purposes. The specific UP-PCR pattern for each isolate can already be used as fingerprints, and as such as a tool for monitoring an isolate of interest in population studies.

So far, we have been able to group the *Trichoderma* and *Gliocladium* isolates into different genomic species by UP-PCR information, and into ITS1 groups. Further work is necessary to confirm the proposed groups by including more isolates, and by obtaining information from analyzing the ITS2-25S rDNA region with restriction enzyme digestion patterns.

We plan to characterize a larger amount of isolates by UP-PCR and ribotyping to establish a collection of reference isolates with known UP-PCR and rDNA polymorphisms. While the UP-PCR polymorphisms have the potential to represent a unique fingerprint for each different isolate, analysis of UP-PCR hybridization data and of rDNA can be used to characterize isolates to appropriate groups and genomic species. Such a collection could be useful for studies of the natural diversity of populations of *Trichoderma* and *Gliocladium* isolates in soil and may help in fast identification or characterization of new isolates.

One of our isolates of *G. roseum*, IK 726, has shown promising biocontrol of seedling diseases caused by *Fusarium culmorum* and *Bipolaris sorokiniana* (Knudsen *et al.*, 1995). From our preliminary results we found that our isolates of *G. roseum* have very similar UP-PCR band patterns, and could be placed in one distinct ITS1 group (Bulat *et al.*, *unpublished data*). The species designation in this case seems to be easy and reliable. We plan to screen isolates of *G. roseum* with a wide range of specially designed universal primers to obtain one or more isolate specific band(s) for isolate IK 726 to develop a PCR-based diagnostic method for this isolate. Such a diagnostic method has a great potential to be used for monitoring this isolate, and in studies of population dynamics in field experiments.

Acknowledgements

We are grateful to Dr. David B. Collinge for critically reading and commenting the manuscript, and thank Dr. Ulf Thrane for providing us with isolates. This research was supported by the Royal Veterinary and Agricultural University and by the grant "Frontiers in Genetics" from Russian Academy of Sciences.

References

Bissett, J. (1984) A revision of the genus *Trichoderma*. I. Section *Longibrachiatum* sect. nov., *Canadian Journal of Botany* **62**, 924-931.

Bissett, J. (1991) A revision of the genus Trichoderma. II. Infrageneric classification. *Canadian Journal of Botany* **69**, 2357-2372.

Bruns, T.D., White, T.J., and Taylor, J.W. (1991) Fungal molecular systematics, *Annual Review of Ecology and Systematics* **22**, 525-564.

Bulat, S.A., and Mironenko, N.V. (1992) Polymorphism of yeast-like fungus *Aureobasidium pullulans* (De Bary) revealed by universally primed polymerase chain reaction: species divergence state, *Genetika* (Russ.) **28**, 19-30.

Bulat, S.A., Mironenko, N.V., Lapteva, M.N., and Strelchenko, P.P. (1994) Polymerase chain reaction with universal primers (UP-PCR) and its application to plant genome analysis, in R. P. Adams *et al.* (eds.), *Conservation of Plant Genes II: Utilization of ancient and modern DNA*, Missouri Botanical Garden, St. Louis **48**, 113-129.

Domsch, K.H., Gams, W., and Anderson, T.H. (1980) *Compendium of Soil Fungi*, Vol. 1. London Academic, pp. 398-377, pp. 794-809.

Ghisalberti, E.L., and Sivasithamparam, K. (1991) Antifungal antibiotics produced by *Trichoderma* spp, *Soil Biology and Biochemistry* **23**, 1011-1020.

Jensen, D.F., and Wolffhechel, H. (1995) The use of fungi, particularly *Trichoderma* spp. and *Gliocladium* spp. to control root rot and damping-off diseases, in H. Hokkanen, and J. M. Lynch (eds.), *Biocontrol Agents: Benefits and Risks*, Cambridge University Press, Cambridge pp. 177-189.

Knudsen, I.M.B., Hockenhull, J., and Jensen, D.F. (1995) Biocontrol of seedling diseases caused by *Fusarium culmorum* and *Bipolaris sorokiniana*: effects of selected fungal antagonists on growth and yield components, *Plant Pathology* **44**, *in press*.

Lübeck, M., and Lübeck, P.S. (1996) PCR - a tool for detection and identification of fungi in soil, *this volume*.

Meyer, R.J. (1991) Mitochondrial DNAs and plasmids as taxonomic characteristics in *Trichoderma viride*, *Applied and Environmental Microbiology* **57**, 2269-76.

Meyer, R.J., and Plaskowitz, J.S. (1989) Scanning electron microscopy of conidia and conidial matrix of *Trichoderma*, *Mycologia* **81**, 312-317.

Meyer, W., Morawetz, R., Börner, T., and Kubicek, C.P. (1992) The use of DNA-fingerprint analysis in the classification of some species of the *Trichoderma* aggregate, *Current Genetics* **21**, 27-30.

Muthumeenakshi, S., Mills, P.R., Brown, A.E., and Seaby, D.A. (1994) Intraspecific molecular variation among *Trichoderma harzianum* isolates colonizing mushroom compost in the British Isles, *Microbiology* **140**, 769-777.

Papavizas, G.C. (1985) *Trichoderma* and *Gliocladium*: biology, ecology, and potential for biocontrol, *Annual Review of Phytopathology* **23**, 23-54.

Rehner, S.A., and Samuels, G.J. (1994) Taxonomy and phylogeny of *Gliocladium* analyzed from nuclear large subunit ribosomal DNA sequences, *Mycological Research* **98**, 625-634.

Rifai, M.A. (1969) A revision of the genus *Trichoderma*, *Mycological Papers* **116**, 1-56.

Stasz, T.E., Nixon, K., Harman, G.E., Weeden, N.F., and Kuter, G.A. (1989) Evaluation of phenetic species and phylogenetic relationships in the genus *Trichoderma* by cladistic analysis of isozyme polymorphism, *Mycologia* **81**, 391-403.

Webster, J., and Lomas, N. (1964) Does *Trichoderma* produce gliotoxin and viridin? *Transactions of the British Mycological Society* **47**, 535-540.

Williams, J.G.K., Kubelik, A.R., Livak, K.J., Rafalski, J.A., and Tingey, S.V. (1990) DNA polymorphisms amplified by arbitrary primers are useful as genetic markers, *Nucleic Acids Research* **18**, 6531-6535.

White, T.J., Bruns, T.D., Lee, S., and Taylor, J.W. (1990) Amplification and direct sequencing of fungal ribosomal RNA genes for phylogenetics, in M. Innis, D.H. Gelfand, J.J. Sninsky, and T.J. White (eds.), *PCR Protocols: a Guide to Methods and Applications*, Academic Press, San Diego, pp. 315-322.

Zimand, G., Valinsky, L., Elad, Y., Chet, I., and Manualis, S. (1994) Use of RAPD procedure for the identification of *Trichoderma* strains, *Mycological Research* **98**, 531-534.

IDENTIFICATION OF NEMATODE-TRAPPING FUNGI OF THE GENUS *ARTHROBOTRYS* USING RFLP ANALYSIS OF PCR-AMPLIFIED rDNA

YVONNE PERSSON, SUSANNE ERLAND & HANS-BÖRJE JANSSON
Department of Microbial Ecology, Lund University, Ecology Building, S-223 62 Lund, Sweden.

1. Introduction

The fungal genus *Arthrobotrys* Corda contains fortyeight various nematode-trapping species. Within the genus fungi that produce traps of the adhesive network type as well as a few species that form constricting rings are found.

The species *Arthrobotrys oligospora* is now divided into three varieties: *Arthrobotrys oligospora* var. *oligospora*, var. *microspra* and var. *sarmatica* . This division has mainly been based on conidial size (van Oorschot, 1985). The complexity within the *Arthrobotrys* genus was stressed by Mankau (1980), who compared the variation and complexity occurring with the species in the genus *Penicillium* and its relatives.

The identification of *Arthrobotrys* spp. and many other nematophagous fungi is based on morphological characters mainly following the keys by Cooke and Godfrey (1964) and van Oorschot (1985). Major problems still exist in the practical daily work of identification of *Arthrobotrys* spp.

With the rapidly developing use of molecular techniques such as the polymerase chain reaction (PCR) in mycology (Foster, *et al.*, 1993) techniques have been developed for rapid and reproducible identification protocols (Erland *et al.*, 1994). Application of such techniques for identification of nematophagous fungi in general is new (Persson *et al.*, 1995), and we believe that these methods can be important tools for rapid and safe identification of such fungi. In the current paper we present results from restriction fragment length polymorphism (RFLP) analysis of the PCR-amplified ITS-region of ribosomal DNA six isolates of *A. oligospora* var. *oligospora* and from four other *Arthrobotrys* species.

2. Material and methods

2.1 FUNGAL ISOLATES

Six isolates of *Arthrobotrys oligospora* var *oligospora*, CBS 115.81, CBS 106.49, CBS 280.86, CBS 289.82, A.R 9113 and A.R. 936, isolated in different parts of the world, were used in this experiment. The other varieties of *A. oligospora* were not available.

D. F. Jensen et al. (eds.), Monitoring Antagonistic Fungi Deliberately Released into the Environment, 129–132.
© 1996 *Kluwer Academic Publishers. Printed in the Netherlands.*

Four different *Arthrobotrys* strains, *A. superba* CBS 107.81, *Arthrobotrys musiformis* CBS 266.83, *A. conoides* CBS 265.83 and *Arthrobotrys robusta* CBS 698.86 were also included.

2.2 DNA EXTRACTION AND AMPLIFICATION

Fungal colonies grown on corn meal agar, diluted 1:10 were used for DNA extraction. The mycelium was scraped off the agar surface and suspended in 400 µl extraction buffer (EB; 200 mM Tris HCl pH 8.5, 250 mM NaCl, 25 mM EDTA, 0.5% SDS)(Raeder & Broda, 1985) containing 10 µl proteinase K (10 mg/ml). To rupture the cell walls the mycelial suspension was put through three cycles of freezing in liquid nitrogen and thawing followed by grinding using sterile plastic pestles (Kontes). The DNA was extracted with phenol:chloroform:isoamylalcohol and subsequencely precipitated with isopropanol (Sambrook *et al.*, 1989). The pellet was dissolved in 30-50 µl TE (10 mM Tris-HCl buffer, pH 8, 1 mM EDTA) and stored at 8°C until use.

The sequences for the primers, ITS1 and ITS4, used for amplification were (5'–TCCGTAGGTGAACCTGCGG) and (5'-TCCTCCGCTTATTGATATGC), respectively (White *et al.*,1990). The reaction components for the PCR were as described by Erland *et al.* (1994): 0.1-10 ng of total DNA, 500 pmol of each of the primers, 20 mM Tris (pH 8.3), 200 µM dNTP, 1.5 mM $MgCl_2$, 50 mM KCl and 2.5 units of Taq polymerase (Promega). To optimise the yield of the PCR reactions different dilutions for each DNA extraction were tried. After an initial denaturation at 95 °C for 3 min., 35 cycles of denaturation at 95°C for 2 min., annealing at 53°C for 25 s and extension at 72°C for 2 min. with a final extension at 72°C for 10 min. Controls without DNA were run in every series to ensure that there was no DNA contamination in reagents or reaction buffers. The degree and specificity of the amplifications were analyzed by 2% agarose gel electrophoresis and ethidium bromide staining (Sambrook *et al.*, 1989). In order to claryfy the differences between the amplificationproducts each was digested with four different restriction enzyme; *Alu*I, *Hin*fI, *Rsa*I and *Taq*I. The digestions were performed according to the manufacturers instructions.

3. Results

The ITS region of all the isolates of nematode-trapping fungi tested was successfully amplified by PCR, using the ITS 1 and ITS 4 primers. The size of the amplification product was 600-700 bp. The PCR products of the isolates were cut by four different restriction enzymes (*Alu*I, *Hin*fI, *Rsa*I and *Taq*I).

Five isolates of *A. oligospora*, originating from different parts of the world, produced identical PCR-RFLP patterns when cut with the four different restriction enzymes. One isolate, CBS 289.82, diverged from the others. This isolate has been reported to capture nematodes directly on its undifferentiated hyphae. This is not the case for any of the other isolates but a similar property has been reported for *A. superba* (Jansson & Nordbring-Hertz, 1981). When the *A. oligospora* var. *oligospora* isolate was compared to an isolate of *A. superba* it showed a restriction pattern similar, but not identical, to *A. superba*.

Comparison between five different *Arthrobotrys* species showed that even though the restriction pattern of one or more enzymes corresponded for two species, at least one enzyme pattern was clearly different. All species differed both from each other and from the *A. oligospora* standard.

4. Discussion

PCR-RFLP of the ITS region of the ribosomal DNA represents a rapid and convenient method for identification of nematode-trapping fungi at the species level. These results are consistent with earlier findings in other groups of fungi (Lee & Taylor, 1992, Huffman *et al.*, 1992, Gardes *et al.*, 1991, Gardes & Bruns, 1993, Erland *et al.*, 1994). We used a set of fungal isolates from fungal culture collections which were identified as correctly as possible on the basis of morphological characteristics.

Five *A. oligospora* var. *oligospora* isolates gave identical restriction patterns even though they all had different geographical origins. One of the isolates, CBS 289.82, differed, however. Due to the fact that this isolate also differs in its nematode-capturing abilities we assume that it belongs to another species. When compared to an *A. superba* isolate it was similar but not identical. So far we have not found a perfect match.

All other species tested had unique restriction patterns. Our results show that amplification and restriction of the ITS fragment using the four restriction enzymes, *Alu*I, *Hin*fI, *Rsa*I and *Taq*I, was sufficient to distinguish between different species of nematophagous fungi. In our study we have used axenic cultures as a DNA source. However, due to the minute amounts of DNA required for amplification by PCR, this method should be possible to develop for direct isolation of fungal DNA from natural habitats, e.g. individual nematodes. This would not only minimise the tedious identification work at the microscope but also eliminate the culturing time.

5. References

Cooke, R.C. & Godfrey, B.E.S. (1964). A key to nematode-destroying fungi. *Transactions of the British Mycological Society,* **47**, 61-74.

Erland, S., Henrion, B., Martin, F., Glower, L.A. & Alexander, I. J.(1994). Identification of the ectomycorrhizal basidiomycete *Tylospora fibrillosa* Donk by RFLP analysis of the PCR-amplified ITS and IGS regions of ribosomal DNA. *New Phytologist* **126**,525-532.

Foster, L.M., Kozak, K.R., Loftus, M.G., Stevens, J.J. & Ross, I.K. (1993). The polymerase chain reaction and its application to filamentous fungi. *Mycological research* **97**, 769-781.

132

Gardes, M. & Bruns T. D. (1993). ITS primers with enhanced specificity for basidiomycetes - application to the identification of mycorrhizae and rusts. *Molecular Ecology* **2,** 113-118.

Gardes, M., White, T. J., Fortin, J. A., Bruns, T. D. & Taylor, J. W. (1991). Identification of indigenous and introduced symbiotic fungi in ectomycorrhizae by amplification of nuclear and mitochondrial ribosomal DNA. *Canadian Journal of Botany* **69,** 180-190.

Huffman, J.L., Molina, F.I. & Jong, S.-C. (1992). Authentication of ATCC strains in the *Saccharomyces cerevisiae* complex by PCR fingerprinting. *Experimental Mycology* **16,** 316-319.

Jansson, H.-B. & Nordbring-Hertz, B (1981). Trap and conidiophore formation in *Artrobotrys superba. Transactions of the British Mycological Socitey* **77,** 203-205.

Lee, S. B. & Taylor, J. W. (1992). Phylogeny of five fungus-like Protoctistan *Phytophtora* species, inferred from the internal transcribed spacers of ribosomal DNA. *Molecular Biology and Evolution* **9,** 636-653.

Mankau, R.(1980). Biological control of nematode pests by natural enemies. *Annual Review of Phytopathology* **18,** 415-440.

Oorschot, van, C.A.N. (1985). Taxonomy of the *Dactylaria* complex. V. A review of *Arthrobotrys* and allied genera. *Studies on Mycology* **26,** 61-96.

Persson, Y., Erland, S. & Jansson, H.-B. (1995). Identification of nematode-trapping fungi using RFLP analysis of the PCR-amplified ITS region of ribosomal DNA (Mycological Research, accepted)

Raeder, U. & Broda, P. (1985). Rapid preparation of DNA from filamentous fungi. *Letters in Applied Microbiology* **1,** 17-20.

Sambrook, J., Fritsch, F. E. & Maniatis, T. (1989). Molecular Cloning: A Laboratory Manual. (Ed. by Cold Spring Harbour Laboratory), Cold Spring Harbor: New York.

White, T. J., Bruns, T., Lee, S. & Taylor, J. (1990). Amplification and direct sequencing of fungal ribosomal RNA genes for phylogenetics. In: *PCR Protocols. A guide to methods and applications.* (ed.M. A. Innis, D. H. Gelfand, J. J. Sninsky & T. J. White), pp 315-322. Academic Press: San Diego.

CLONING AND CHARACTERIZATION OF *TRICHODERMA HARZIANUM* GENES INDUCED DURING GROWTH ON *RHIZOCTONIA SOLANI* CELL WALLS

VALERIE VASSEUR*, MARC VAN MONTAGU AND GUSTAVO H. GOLDMAN
Laboratorium voor Genetica, Universiteit Gent, K.L. Ledeganckstraat 35, B-9000 Gent, Belgium.

* *Present address : ESMISAB, Laboratoire de Microbiologie et Biochimie, Technopôle Brest-Iroise, F-29280 Plouzané, France.*

1. Abstract

Trichoderma harzianum is a biocontrol agent that attacks a large variety of phytopathogenic fungi. In an attempt to better understand the mycoparasitism process, genes which are specifically expressed by *T. harzianum* during growth on cell walls of *Rhizoctonia solani* have been studied. It is carried out a differential screening of an induced cDNA library. Two cDNA clones have been isolated and characterized. One of these clones corresponds to a gene that encodes a presumptive amino acid permease. The other encodes a novel protein abundantly expressed.
The use of this methodology for identifying and characterizing mycoparasitism-related genes should provide more biochemical informations about mycoparasitism process, and provide specific markers to follow the biological interaction.

2. Introduction

Trichoderma harzianum is a biocontrol agent that attacks a large variety of phytopathogenic fungi responsible for major crop diseases. In soil, *T. harzianum* strains have been used as antagonists against several plant-pathogenic fungi, *e.g. Rhizoctonia solani, Botrytis cinerea* and *Sclerotium rolfsii* (1). A great deal of information about the factors involved in the antagonistic properties of *T. harzianum* have been identified. These include volatile and non-volatile antibiotics and hydrolytic enzymes such as chitinases, glucanases and proteases. Recently, two mycoparasitism-related genes from *T. harzianum*, encoding an alkaline proteinase (2) and an endochitinase (3) have been isolated and characterized.
When *T. harzianum* grows on *R. solani* cell walls as sole carbon source, induction and/or derepression of genes involved in mycoparasitism has been found to occur.
Additionally, glucose represses the activity and/or expression of hydrolytic enzymes important for the degradation of the cell walls (2, 4).
Our approach to identify factors determining mycoparasitic abilities of *T. harzianum* was to isolate genes of this species that are specially expressed during growth on *R. solani* cell walls. In an attempt to identify these genes, it is carried out differential screening of

D. F. Jensen et al. (eds.), Monitoring Antagonistic Fungi Deliberately Released into the Environment, 133–137.

134

an induced cDNA library. Two cDNA clones were isolated. One corresponds to a gene encoding an amino acid permease and the other corresponds to a gene encoding a novel polypeptide.

3. Results

3.1. DIFFERENTIAL SCREENING

Specific cDNA clones from a *T. harzianum* cell wall-grown cDNA library (constructed using poly(A)$^+$ -RNA obtained from *T. harzianum* grown on cell walls as sole carbon source) were isolated by differential screening against probes from *T. harzianum* grown on glucose or on cell walls of *R. solani*. The initial screening resulted in the identification of 65 clones, and a secondary screening in 15 cDNA clones that hybridized only to cDNA probes derived from *T. harzianum* grown on cell walls. For further analysis, the clones which hybridized only on Northern analysis experiments with mRNAs isolated from cell-wall-grown *T. harzianum* were chosen (see below). At the end of the screening, 6 different clones which are expressed specifically during growth on cell walls of *R. solani* were obtained.
Two of these cDNA clones, INDA1 and INDC11, were isolated and analysed.

3.2. CHARACTERIZATION OF THE INDA1 AND INDC11 CDNA CLONES

3.2.1. Northern Analyses
To clarify the expression of *inda1* and *indc11* genes, mRNAs were isolated from *T. harzianum* grown on cell walls or glucose and Northern analyses were carried out (Figure 1).

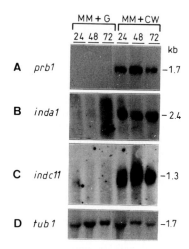

Figure 1. Northern blot of the mRNAs obtained from *T. harzianum* grown on minimal medium plus glucose (MM+G) or *R. solani* cell walls (MM+CW). Transcript levels of *prb1* (A), *inda1* (B), *indc11* (C), *tub1* (D). *tub1* is a gene encoding a β-tubulin from *T. viride*. Mycelia were harvested 24, 48, 72 hours after transfer into MM+G or MM+CW.

One gene expressed during mycoparasitism by *T. harzianum* is the *prb1* gene, that encodes an alkaline proteinase (2). Its expression was repressed in the presence of glucose and increased during cell-wall-grown *T. harzianum* (Figure 1A).
Both genes *inda1* and *indc11* did not hybridize to mRNAs isolated during growth of *T. harzianum* in the presence of glucose. They hybridized only with mRNA obtained from *T. harzianum* grown on cell walls. The *inda1* gene specifies a single transcript of about 2.4 kb whereas the *indc11* gene specifies a single transcript of about 1.3 kb (Figures 1B and 1C).
β-Tubulin mRNA expression was constant during growth on glucose and cell walls (Figure 1D).

3.2.2. DNA Blot Hybridizations

DNA blot hybridizations are performed with genomic DNA prepared from *T. harzianum*. After digestion of the genomic DNA with appropriate restriction enzymes, few fragments could be identified in the restriction map of the clones hybridized to the INDA1 cDNA and INDC11 cDNA respectively, indicating that *inda1* and *indc11* are most probably single-copy genes (Figure 2).

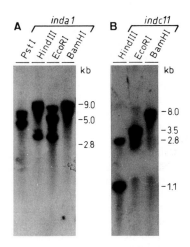

Figure 2. Southern blot of *T. harzianum* genomic DNA. DNA was digested with restriction endonucleases *Bam*HI, *Eco*RI, *Hind*III, or *Pst*I. Blots were probed with a ^{32}P-labelled 2.1-kb fragment from INDA1 cDNA clone (A) or a 1.2-kb fragment from INDC11 cDNA clone (B).

3.2.3. Sequences of the cDNA Clones

The *inda1* and *indc11* genes have been sequenced.
The INDA1 cDNA clone contained an insert of 2.1 kb. INDA1 cDNA clone corresponds to a gene that encodes a putative amino acid permease. The codon usage of the *inda1* gene resemble that of other *Trichoderma* genes. The *inda1* open reading frame encodes a protein of 574 amino acid residues with a predicted molecular mass of 62853 Da and a calculated pI value of 7.81. The INDA1 amino acid sequence showed high similarity with other sequenced amino acid permeases from different organisms as *Saccharomyces cerevisiae* and *Aspergillus nidulans*.

The INDC11 cDNA clone contains an insert of approximately 1.2 kb. INDC11 cDNA clone corresponds to a gene that encodes a novel polypeptide abundantly expressed.
The predicted protein is 340 amino acids long, having a calculated molecular mass of 37010 Da and a predicted pI value of 8.25.
The *indc11* nucleotide and amino acid sequences do not display high homology with any sequence deposited in the data banks. The typical N-terminal sequence resembling signal sequences for excreted proteins was not found, suggesting that the INDC11 is not a secreted protein.

4. Conclusion

We have isolated and characterized two cDNA clones that correspond to genes that encode novel putative mycoparasitism-related proteins of *T. harzianum*. These genes are expressed during growth on *R. solani* cell walls.
Further investigations are required to determine the actual role of the genes in the mycoparasitic interaction. The different steps involved in the mycoparasitism process have been described (1). Cell wall degradation by hydrolytic enzymes is probably involved in the process of mycoparasite penetration. These enzymes have been detected when *T. harzianum* was grown on *R. solani* cell walls as sole carbon source (4). An alkaline proteinase specifically induced by *R. solani* cell walls has been purified and the corresponding gene cloned (2). We suggested that the PRB1 was involved in the degradation of the phytopathogen cell walls and possibly in the degradation of proteins released after lysis of the host. Thus, the mycoparasite could utilize amino acids derived from proteins hydrolysis as carbon and/or nitrogen sources. Consequently, the uptake of these amino acids will be of importance for the nutrition of the mycoparasite. The coordinated transcription of the *prb1* and *inda1* genes fits this model, but formal proof in this hypothesis must await additional studies.
    The use of our methodology should provide more biochemical informations about mycoparasitism as well as provide specific genetic markers to follow the biological interaction. Further steps in this direction would be knock-out of the *inda1* and *indc11* genes in order to prove their essentiality during the process of mycoparasitism.
    Investigations will be carried out into the development of better biocontrol strains of *Trichoderma* spp. The fact that *inda1* and *indc11* genes are expressed during growth on *R. solani* cell walls suggests that their promoters could be useful for many types of genetic engineering. Thus, the fusion of genes, encoding chitinase or glucanase, under the control of these promoters could have great potential for use in producing transgenic *Trichoderma* strains with enhanced biocontrol activity.

5. References

1. Chet, I. (1987) *Trichoderma*-application, mode of action, and potential as a biocontrol agent of soilborne plant pathogenic fungi, in I. Chet (ed.), *Pests-Biological Control*, (Series in Ecological and Applied Microbiology), John Wiley, New York, pp. 137-160.

2. Geremia, R., Goldman, G.H., Jacobs, D., Ardiles, W., Vila, S.B., Van Montagu, M. and Herrera-Estrella, A. (1993) Molecular characterization of the proteinase-encoding gene, *prb1*, related to mycoparasitism by *Trichoderma harzianum, Molecular Microbiology* **8**, 603-613.

3. Hayes, C.K., Klemsdal, S., Lorito, M., Di Pietro, A., Peterbauer, C., Nakas, J.P., Tronsmo, A. and Harman, G.E. (1994) Isolation and sequence of an endochitinase-encoding gene from a cDNA library of *Trichoderma harzianum, Gene* **138**, 143-148.

4. Geremia, R., Jacobs, D., Goldman, G.H., Van Montagu, M. and Herrera-Estrella, A. (1991) Induction and secretion of hydrolytic enzymes by the biocontrol agent *Trichoderma harzianum*, in A.B.R. Beemster, G.J. Bollen, M. Gerlagh, M.A. Ruissen, B. Shippers and A. Tempel (eds.), *Biotic Interactions and Soil-Borne Diseases*, (Developments in Agricultural and Managed-Forest Ecology Series, Vol. 23), Elsevier, Amsterdam, pp. 181-186.

DEVELOPMENT OF IMMUNOASSAYS FOR THE DETECTION AND QUANTIFICATION OF FUNGI

FRANCES M. DEWEY
*Department of Plant Sciences, University of Oxford, South Parks Rd.,
Oxford, OX1 3RB*

1. Introduction

The development of immunoassays for the detection and quantification of fungi has
been slow; this has been mainly because of the difficulties encountered in raising
antisera to fungi that are species-specific. Antisera raised to mycelial fragments,
extracts from lyophilized mycelia, surface washings of solid cultures or culture filtrates
generally cross-react with both related and unrelated fungi and host tissues or their
extracts when tested by enzyme linked imunosorbent assays (ELISA) or by
immunofluorescence (IMF)(Dewey, 1989a, Mohan, 1988; Xia *et al.*, 1992). Fungi
share many common molecules and binding sites (epitopes) with each other and with
their hosts i.e. common antigens (Devay & Adler, 1976; Chakraborty & Sinha, 1994).
Attempts to improve specificities, by using, as immunogens, purified fractions from
fungi such as hyphal walls, have been only partially successful. However, Notermans
et al .(1987), working with fungi causing spoilage of foods, have raised genus-specific
antisera by using the soluble carbohydrates extracted from supernatants of fungi grown
in liquid culture as the immunogen. Removal of cross-reactive antibodies by cross-
absorption with antigens from related species is not satisfactory because the titre is
usually severely lowered (Gerik *et al.*, 1987).

2. Raising monoclonal antibodies to fungi.

The advent of hybridoma technology (Kohler and Milstein, 1975) has made it possible
to raise and select clones of hybridoma cells that secrete highly specific antibodies (i.e.
monoclonal antibodies [MAbs]). Genus- and species-specific MAbs have now been
raised to a number of fungi (Schots, *et al.*, 1994, Hardham *et al.*, 1986) but relatively
few of these MAbs are available commercially. Although the costs and time involved in
raising monoclonal antibodies are high ,they are generally less than those involved in
developing nucleic acid detection probes. Once developed, specific immunoassays have
many advantages; they can be replicated easily, quantified, and made 'user friendly' for
on-site field studies. Furthermore, they can be adapted for determining sequential
growth of pathogens in soils (see Thornton *et al.,.ibid.*). An outline of the processes
involved in raising specific-MAbs to fungi is presented in Figure 1; for practical details
see Harlow and Lane (1988) and Dewey *et al.*, (1989a, 1989b). Once hybridoma cell
lines secreting specific MAbs have been raised, they can be preserved indefinitely by
freezing slowly to - 80^0C in a mixture of foetal bovine serum and dimethyl sulfoxide

D. F. Jensen et al. (eds.), Monitoring Antagonistic Fungi Deliberately Released into the Environment, 139–146.
© 1996 *Kluwer Academic Publishers. Printed in the Netherlands.*

140

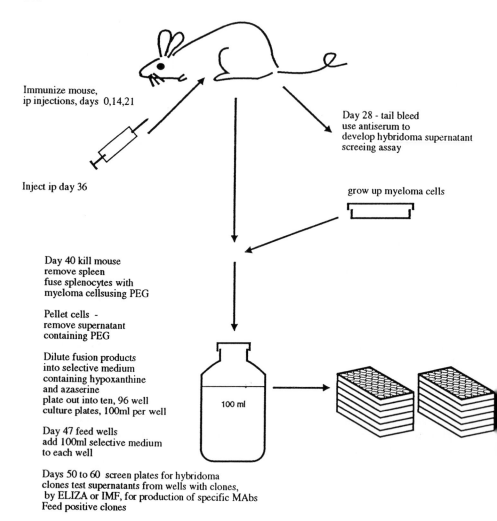

Immunize mouse,
ip injections, days 0,14,21

Day 28 - tail bleed
use antiserum to
develop hybridoma supernatant
screeing assay

Inject ip day 36

grow up myeloma cells

Day 40 kill mouse
remove spleen
fuse splenocytes with
myeloma cellsusing PEG

Pellet cells -
remove supernatant
containing PEG

Dilute fusion products
into selective medium
containing hypoxanthine
and azaserine
plate out into ten, 96 well
culture plates, 100ml per well

100 ml

Day 47 feed wells
add 100ml selective medium
to each well

Days 50 to 60 screen plates for hybridoma
clones test supernatants from wells with clones,
 by ELIZA or IMF, for production of specific MAbs
Feed positive clones

Days 52 to 62
re-test positive wells

Day 53 to 73 subculture positive cell lines into 12, then 6 well plates
further test positive cell lines against related and unrelated fungi and hosts
clone positive cell lines

Days 63 to 93 repeat cloning and testing steps

Approx. day 93 onwards - bulk up selected cell lines in petri dishes and flasks and feeze down .

Figure 1. Outline of steps inolved in raising monoclonal antibodies to specific fungi

and then stored in liquid nitrogen. Frozen cell lines, thawed quickly, can be easily grown again in bulk for further production of the specific MAbs.

2.1 IMMUNOGENS

The process of raising fungal-specific monoclonal antibodies is still relatively inefficient in that, whatever fungal fragment or extract is used as the immunogen, a large percentage of the hybridoma clones produced secrete non-specific MAbs. No one immunogen appears to be more specific than another. We commonly use extracellular antigens obtained by washing the surfaces of solid slant cultures with phosphate buffered saline (Dewey et al., 1989b, 1990). The great advantage of this method is that it is simple, requires little preparation and no processing. The washings can be injected directly into the mouse without freeze-drying. If necessary, such washings can be frozen and thawed without the apparent loss of antigenicity.

Recently, we have managed to improve the efficacy of our immunizations by using a co-immunisation techniques first developed by Barclay and Smith (1986) raising MAbs to bacteria. Mice are injected with antigens from the target fungus together with antibodies previously raised to a related fungus or extracts of host material. This blocks the formation of cross-reactive antibodies thereby reducing the number of hybridoma cell lines secreting non-specific MAbs. Consequently, the time spent on doing secondary and tertiary screening assays is reduced considerably. We found this technique to be particularly helpful in raising specific MAbs to 3 different species of aquatic hyphomycetes that are the primary colonisers of leaves in streams. Antiserum raised to extracts from uninfected leaf material was used as the blocker (Bermingham *et al.*, 1995, Bermingham personal comm.).

2.2 SELECTION OF HYBRIDOMA CELL LINES

When raising monoclonal antibodies for the purposes of detection and/or quantification it is important to use the same assay format for screening and selecting the hybridoma cell lines as that for which the antibodies are finally intended. This is because although many MAbs will recognize the antigen and work well in one format they will not work in other formats. In our experience fungal MAbs will often recognize the immunogen by ELISA using antigen coated wells but they will not recognize the antigen by Western Blotting or by immunofluorescence.

3. Types of immunoassays

Several different assay formats exist but the most common are the ELISA-based systems. These assays have many advantages; they are highly sensitive and can be easily replicated, automated and quantified. The only disadvantage is that they require laboratory facilities.

3.1 ELISA

Most of the ELISA tests developed for fungi are, with the exception of the commercial assays, all simple indirect assays in which the micro-titre wells are directly coated with the test sample (Fig 2). Fungal antigens, particularly glycoproteins, which appear to be the immunodominant molecules, bind strongly to certain makes of micro-titre wells; we find 1 x 12 well miccrostrips produced by Labsystems, Finland, to be particularly effective. Once the wells have been coated with fungal carbohydrates or glycoproteins, washed and dried, they can be stored dry at 4^0C for several years (Dewey, 1992) . The disadvantage of using antigen-coated wells is the relatively long binding step, 5 hours or more (generally an overnight step) needed to ensure maximum binding of the antigens to the wells but the simplicity of such assays is attractive. Fungal carbohydrates and most glycoproteins are heat stable, which means than many of the cross-reactive proteinaceous antigens can be precipitated from the test sample by heat treatment before the antigen mixture is used to coat wells. For example, heat treatment of soil extracts has proved useful in the detection of antigens from *Rhizoctonia solani* (Thornton *et al., ibid.*). We have found that assays involving antigen-coated wells work particularly well where the fungus is present on or near the surface of the infected tissue such as rice grains (Dewey *et al.*, 1989b,1990) and where passive release by overnight soaking is sufficient to enable detection at very low infection levels. Such methods could prove useful in the detection of fungal spores on petals or seed-borne pathogens that are present at or near the surface of the seed coats.

Layer					
4			Sa-E		Sa-E
3		**Ab-E**	**Ab-B**	SpA-E	SpA-B
2	Ab-E	Ab	Ab	Ab	Ab
1	Ag	Ag	Ag	Ag	Ag

Symbols

Ab	Primary Antibody
Ab	**Secondary antibody** (that recognzies the primary Ab)
Ag	Antigen
E	Enzyme
SpA	*Staphylococcus aureus* protein A
B	Biotin
Sa	Streptavidin
-	Indicates conjugation

Figure 2. Various ELISA formats using antigen coated wells

Most of the commercial enzyme immunoassays involve double-antibody -sandwich tests in which one of the antibodies is generally a specific MAb and the other polyclonal antisera (Fig. 3). Sensitivity in both assay types can be improved by using amplification systems such as biotinylated secondary antibodies followed by streptavidin enzyme conjugates or an NADH-NADH enzyme cycling system such as that sold by DAKO, (Denmark ,AMPAK^{TM}). We have found that biological amplification methods are particularly helpful for detection of grain or seed-borne fungi; test samples are soaked in buffer or water, in microtitre wells, overnight at room temperature instead of 4^0C. This allows fungal propagules to germinate and secrete antigens that bind to microtitre wells.

```
Layer
4                        Ab-E           SpA-E
3         Ab-E          Ab             Ab
2         Ag            Ag             Ag
1         Ab            Ab             F(ab)2
```

```
Symbols
Ab        Primary Antibody
Ab        Secondary antibody
Ag        Antigen
E         Enzyme
SpA       Staphylococcus aureus Protein A
F (ab')2  Antibody fragment
-         Indicates conjugation
```

Figure. 3 Selected variations in the design of double-antibody-sandwich (DAS)-ELISAs

In general IgG antibodies present in antiserum recognize taxonomically more specific antigens than do IgM antibodies but it must be stressed that, at the monoclonal level ,we have found that some fungal IgM MAbs are highly specific; they recognize carbohydrate epitopes that frequently occur as repeat epitopes on one molecule and are thus highly sensitive allowing detection levels in the nanogram to fentogram range (see Dewey *et al.*, 1994).

Magnetic beads provide a promising alternative solid support system to micro-titre wells for capturing fungal antigens. Immunoassays incorporating magnetized beads have been used for some time in the fields of medicine and plant virology but their use in the diagnosis of fungi has yet to be fully exploited. Thornton *et al.*, *(ibid)* have developed a magnetic bead immunoassay for the detection and enumeration of propagules of *R. solani* in complex environments.

3.2 DOT-BLOT, DIP-STICK AND TISSUE PRINTING ASSAYS

The mechanism of the dot-blot or dip-stick assay is essentially the same as that in ELISA tests. Some test systems use nitrocellulose membranes or nitrocellulose-coated tags or cards and others use polyvinylidene difluoride (Immobilon P, Millipore). The reporter conjugate in these assays is generally an enzyme conjugate but gold conjugates, which can be silver enhanced, have been used by a number of workers and are thought to be more sensitive (Dewey *et al.*, 1989, Cahill and Hardham, 1994).

Tissue printing or squash blot systems hold considerable promise. They have been used successfully for the detection of an endophyte in grasses (Gwinn *et al.*, 1991). Airie et al., (1993) have developed a combined culture/immunoblot assay for the detection of pathogens in soils in which fungal propagules in the soil grow during the incubation period and release antigens which diffuse through the gel onto the membrane. These membrane bound antigens are then immunolabelled. This method is not dissimilar from that used by Gleason et al., (1987), for the detection of *Phomopsis longicolla* in soybean seed.

3.3 IMMUNOFLUORESCENCE ASSAYS

Imunofluorescence techniques are very useful for visualization of fungal antigens but they are never likely to become widely used for mass screening because the involve microscopy and a UV light source. Nonetheless, they may prove useful in epidemiology for the enumeration of airborne fungal spores settling on leaves or petals (Salinas and Schots, 1994) or trapped on tapes. (Dewey and Kennedy, unpublished) Using a modified tape in a Berkhard spore trap we have been able to collect and distinguish spores of *Botrytis cinerea* from other airborne spores using the genus specific MAb BC-KH4 (Fig 4).

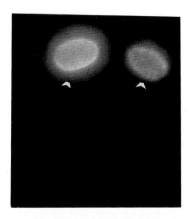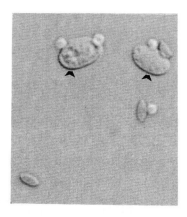

Figure 4. Spores of *Botrytis cinerea* (arrows) and other airborne spores trapped on a modified tape and immunolabled with the Botrytis MAb, BC-KH4, viewed a) with bright light and b) with UV light. Note only Botrytis spores are immunolabelled and visible under UV.

4. Extraction of antigens from plant tissues and soils

Many workers experience difficulties in extracting fungal antigens from infected tissues and soils. Buffers used for extraction of antigens vary; phosphate buffered saline (PBS, pH 7.2) is commonly used for antigen coating of wells where carbohydrate epitopes are involved and bicarbonate buffers (pH 9.6) for systems employing antibodies recognising proteinaceous epitopes. Many buffer systems include azide or thimerosol to prevent growth of contaminant microorganisms. PVP is commonly added to sample buffers to reduce interference from plant phenolic compounds. In sandwich type immunoassay systems antigen extraction buffers generally include 0.05% Tween 20 and reconstituted non-fat dried milk, casein or BSA blockers to reduce non-specific binding.

5. Use of immunoassays to quantify specific fungi

The immunoassay system which lends itself most easily to the determination of the biomass of specific fungi in mixed environments is the ELISA system; a linear relationship can easily be established between the antigen concentration of the fungus and absorbance values (Dewey *et al.*, 1994; Beckman *et al.*, 1994). However, in all immunological quantitative assays it is most important to include a set of standards. Absorbance values for any ELISA test will vary from day to day but the values from well to well within one plate are comparable and therefore the results of test samples are always relative to the set of standards. Initially, it is important to run a dilution series of the test samples because although the results are comparable with a given range it is important to establish that the concentration of the test material is not outside the assay range i.e. too low or too high. The absorbance values for most fungal MAb ELISA tests are linear within the nanogram range. Frequently a dilution series of PBS extracts from freeze dried mycelium are used as standards.

6. References

Arie, T., Hayashi, S., Schimazaki, K., Yoneyama, I. and Yamaguchi. (1993) Novel diagnosis of *Fusarium* infestation of seedlings and soils by immunoassay methods, in *Proceedings of the Sixth International Congress of Plant Pathology*, Montreal, Canada, July-August 1993. Abstract 2.1.15., p. 40.

Barclay, S.L. and Smith, A.M. (1986) Rapid isolation of monoclonal antibodies specific for cell surface antigens, *Proceedings of the National Academy of Sciences USA* **83,** 4336-4340.

Beckman, K.B., Harrison, J.G. and Ingram, D.S. (1994) Optimisation of a polyclonal enzyme linked immunosorbent assay (ELISA) of fungal biomass for use in studies oof plant defence. *Physiological and Molecular Plant pathology* **44,** 19-32.

Bermingham, S., Dewey, F.M. and Maltby, L. (1995) Development of a monoclonal antibody based immunoassay for the detection and quantification of *Anguillospora longissima* colonising leaf material, *Applied and Environonmental Microbiology* (in press)

Bossi, R. and Dewey, F.M. (1992). Development of a monoclonal antibody-based immunodetection assay for *Botrytis cinerea*, *Plant Pathology* **41,** 472-482.

Cahill, D.M. and Hardham, A.R. (1994a) Exploitation of zoospore taxis in the development of a novel dipstick immunoassay for the detection of *Phytophthora cinnamomi*, *Phytopathology* **84,**193-200.

Chakraborty, B.N. and Sinha, A. (1994) Detection and cellular location of cross-reactive antigens shared by *Camellia sinensis* and *Bipolaris carbonum*, *Physiological and Molecular Plant pathology* **44,** 403-416.

Devay, J.E. and Adler, H.E. (1976) Antigens common to hosts and parasites. *Annual Review of Microbiology* **30,** 147-168.

Dewey, F.M., Munday, C.J. and Brasier, C.M. (1989a) Monoclonal antibodies to specific components on the Dutch elm pathogen *Ophiostoma ulmi*, *Plant Pathology* **38,** 9-20.

Dewey, F.M., McDonald, M.M. and Phillips, S.I. (1989b) Development of Monoclonal-antibody-ELISA, -DOT-BLOT and -DIP-STICK Immunoassays for *Humicola lanuginosa* in Rice, *Journal of General Microbiology* **135,** 361-374.

Dewey, F.M., McDonald, M.M., Phillips, S.I. and Priestley, R.A. (1990) Development of monoclonal-antibody-ELISA and -DIP-STICK immunoassays for *Penicillium islandicum* in rice grains, *Journal of General Microbiology* **136,** 753-760.

Dewey, F.M. (1992) Detection of plant invading fungi by monoclonal antibodies, in J.M. Duncan and L. Torrance (eds), *Techniques for the Rapid Detection of Plant Pathogens,* Blackwell Scientific Publications Ltd., Oxford, pp. 47-62.

Dewey, F.M., Twiddy, D.R., Phillips, S.I. Gross, M.T. and Wareing, P.W. (1992) Development of a quantitative monoclonal antibody-based immunoassay for *Humicola lanuginosa* on rice grains and comparison with conventional assays, *Food and Agricultural Immunology* **4,** 153-167.

Gerik, J.S., Lommel, S.A., Huisman, O.C. (1987) A specific serological staining procedure for *Verticillium dahliae* in cotton root tissue, *Phytopathology* **77,** 261-266.

Gleason, M.L., Ghabrial, S.A. and Ferris, R.S. (1987) Serological detection of *Phomopsis longicolla* in soybean seeds, *Phytopathology* **77,** 371-375.

Gwinn, K.D., Collins-Shephard, M.H. and Reddick, B.B. (1991) Tissue print-immunoblot, an accurate method for the detection of *Acremonium coenophialum* in tall fescue, *Phytopatholgy* **81** (7), 747-748.

Hardham, A.R., Suzaki, E. and Perkin, J.L. (1986) Monoclonal antibodies to isolate-, species-,and genus-specific components on the surface of zoospores and cysts of the fungus *Phytophthora cinnamomi*, *Canadian Journal of Botany* **64,** 3 11-321.

Harlow, E. and lane, D. (1988) *Antibodies - a laboratory manual,* Cold Spring harbour, pp139-319.

Kohler, G. and Milstein, C. (1975) Continuoous culture of fused cells secreting antibody of pre-defined sspecificity, *Nature* **256,** 495-497.

Mohan, S.B. (1988) Evaluation of antisera raised against *Phytophthora fragariae* for detecting the red core disease of strawberries by enzyme-linked immunosorbent assay (ELISA) *Plant Pathology* **38,** 352-363.

Notermans, S., Wieten, G., Engel, H.W.B., Rombouts, R.M., Hoogerhout, P. and Van Boom, J.H. (1987) Purification and properties of extracellular polysaccharide (EPS) antigens produced by different mould species, *Journal of Applied Bacteriology* **62,** 157-166.

146

Salinas, J., and Schots, A. (1994) Monoclonal Antibodies-based test for detection of cconidia of *Botrytis cinerea* on Cut Flowers, *Phytopathology* **84**, 351-356.

Schots., Dewey, F.M. and Oliver, R. (1994) *Modern Assays for Plant Pathogenic Fungi; Identification, Detection and Quantification,* CAB International, Wallingford, U.K.

Thornton, C.R., Dewey, F.M. and Gilligan, C.A. (1994) Development of a monoclonal antibody-based enzyme-linked immunosorbent assay for the detection of live propagules of *Trichoderma harzianum* in a peat/ bran medium, *Soil Biology and Biochemistry* **26**, 909-920..

Thornton, C.R., Dewey, F.M. and Gilligan, C.A. (1993b) Development of monoclonal antibody-based immunological assays for the detection of live propagules of *Rhizoctonia solani* in soil, *Plant Pathology* **42**, 763-773.

Xia, J.Q., Lee, F.N., Scott, H.A. and Raymond, L.R. (1992) Development of monoclonal antibodies specific for *Pyricularia grisea*, the rice blast pathogen, *Mycological Research* **96**, 867-873.

DEVELOPMENT OF MONOCLONAL ANTIBODY-BASED IMMUNOASSAYS FOR THE QUANTIFICATION OF RHIZOCTONIA SOLANI AND TRICHODERMA HARZIANUM IN SOIL

CHRISTOPHER R. THORNTON

DEPARTMENT OF PLANT SCIENCES,
CAMBRIDGE UNIVERSITY,
DOWNING STREET,
CAMBRIDGE.
CB2 3EA.

1. INTRODUCTION

Reliable quantification of soil-borne organisms is a major limiting factor in the experimental investigation of microbial interactions in soil as well as in the testing of ecological and epidemiological theory for these organisms (Gilligan, 1983). Experimental progress in this field has been severely limited by the lack of suitable methods to detect and quantify specific components of fungal biomass in soil samples (Ogoshi, 1987). Recent work on the dynamics of soil-borne diseases and microbial antagonists has depended, therefore, upon imprecise and arguably biased estimates of inoculum to test models for, biological control (Gilligan, 1990). Consequently, there is a need for relatively quick and specific assays, such as immunological assays, to detect and quantify specific fungi in soil samples. The aim of this publication is to report the development of monoclonal antibody-based immunological assays for the detection and quantification of *Rhizoctonia solani* and *Trichoderma harzianum* in soil.

2. MONOCLONAL ANTIBODIES

2.1. RHIZOCTONIA SOLANI

Murine monoclonal antibodies (MAbs) were raised against an anastomosis group (AG) 4 isolate of *Rhizoctonia solani* (Thornton *et al.*, 1993). Mice were immunized using either phosphate buffered saline (PBS) suspensions of lyophilized mycelium plus Quil A adjuvant, or with a solubilized acetone precipitate prepared from cell-free surface washings from solid slant cultures. Polyclonal antisera (PAbs) were raised in rabbits using PBS suspensions of lyophilized mycelium and Quil A adjuvant.

Two of the mice hybridoma cell lines raised, EH2 and EE1, produced immunoglobulins of the classes IgA and IgM respectively that were species-specific recognizing, by enzyme-linked immunosorbent assay (ELISA), antigens from isolates of *R. solani* belonging to AG's 1, 2-1, 2-2, 3, 4, 5, 6, 7, and 8. They did not cross-react, by ELISA, with antigens from other related or unrelated soil-borne fungi including *R. cerealis*

147

D. F. Jensen et al. (eds.), Monitoring Antagonistic Fungi Deliberately Released into the Environment, 147–153.

and *R. carotae*. Characterization of the AG4 antigen by heat, protease and periodate treatment showed that the IgA and IgM MAbs recognized protein and carbohydrate epitopes respectively. These form part of a larger extracellular glycoprotein molecule.

2.2. TRICHODERMA HARZIANUM

2.2.1. *Mycelium-specific monoclonal antibodies*

Murine MAbs were raised against mycelium of a *Trichoderma harzianum* isolate T95 (Thornton *et al.*, 1994). A hybridoma cell line, HD3, produced MAbs of the immunoglobulin sub-class IgG_{2a} that were mycelium-specific recognizing, by ELISA, antigens from a number of *Trichoderma* species and species from the related genus *Hypocrea* but not from a range of other soil-borne fungi. These MAbs did not recognize antigens from phialoconidia or chlamydospores of T95. Characterization of the T95 mycelial antigen and binding on Western blots showed that the HD3 MAbs recognized a protein epitope of *ca* M_r 18,500 which forms part of a larger glycoprotein molecule.

2.2.2. *Phialoconidia-specific monoclonal antibodies*

Murine MAbs were raised against phialoconidia of *T. harzianum* isolate T95 (Thornton and Dewey, 1995). A hybridoma cell line, GH5, produced MAbs of the immunoglobulin class IgM. GH5 MAbs were genus-specific recognizing, by ELISA, antigens from a number of *T. harzianum* and *T. viride* isolates. Characterization of the T95 conidial antigens showed that the HD3 MAbs recognized a protein epitope.

3. ENZYME IMMUNOASSAYS

3.1. RHIZOCTONIA SOLANI

The IgA and IgM MAbs produced by the cell lines EH2 and EE1 were used to develop a number of immunological assays for the detection and quantification of live mycelium of *R. solani* in soil. The assays and their appropriate applications are summarized in TABLE 1. The methodology of each assay can be found in Thornton *et al.* (1993) with the exception of the immuno-blotting method (see Thornton *et al.*, 1995a).

TABLE 1. Detection and quantification of *R. solani* in soil: immunoassay formats and their appropriate applications

Immunoassay	Application
Immunofluorescence	*In situ* visualization/
Immuno-blotting	differentiation
DIP-STICK	'On-site detection'
Plate trapped antigen (PTA)-ELISA	Quantification
Immuno-magnetic bead assay	Detection and quantification

3.1.1. *Immunofluorescence*

This method markedly improves the accuracy of fungal isolation techniques using selective media and allows the differentiation between colonies of the target fungus, *R. solani,* and contaminant saprophytic fungi, growing in Petri-dish culture.

3.1.2. *Immuno-blotting*

This is a non-destructive immunological method that allows the quantitative visualization of *R. solani*, in soil, *in situ.* Quantification can be achieved in rhizosphere and bulk soil in the presence and absence of *T. harzianum* and a susceptible host, for example radish (*Raphinus sativus*). The immuno-blotting system comprises infested soil sandwiched between nylon micro-filament filter cloth and glass plates. At successive times the plates are removed and overlaid with PVDF transfer membrane. Fungal antigens that bind to the membrane are visualized with successive incubations in EE1 IgM MAb supernatant, secondary anti-mouse IgM gold conjugate and silver enhancement. This technique is non-destructive, *cf* buried slide techniques and immunofluorescence, and enables dynamical observations of the local saprophytic and parasitic growth of different isolates of *R. solani* (see *Figure* 1., with further details in Thornton *et al.,* 1995a).

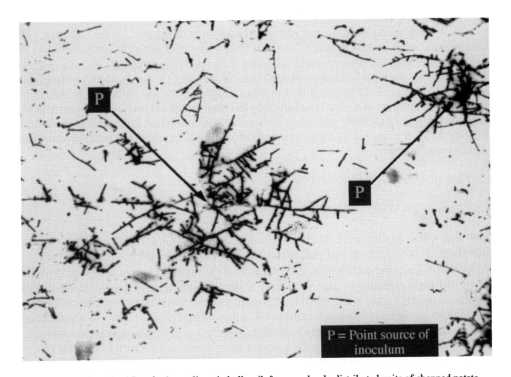

Figure 1. Growth of *R. solani* mycelium, in bulk soil, from randomly distributed units of chopped potato inoculum, visualized by detection of membrane-bound antigen using the immuno-blotting technique. Radial growth of mycelium from point sources of primary inoculum (arrowed) are clearly visible.

3.1.3. DIP-STICK and plate trapped antigen (PTA)-ELISA

Both assays involve the immobilization of fungal antigen(s) on to a solid phase (PVDF membrane or PVC microtitre wells respectively) which often requires an overnight antigen incubation step. Subsequent visualization of bound antigen is achieved by ELISA. While PTA-ELISA systems are extremely sensitive and are readily quantified they are, by their nature, restricted to the laboratory. In contrast, the DIP-STICK assay is more adaptable and could prove extremely useful for the 'on-site' detection of soil-borne fungal pathogens.

3.1.4. Immuno-magnetic bead assay

One of the most significant rate determining step in the development of sensitive immunoassays for the detection and quantification of soil-borne fungi is the extraction of their antigens. Most assays require lengthy extraction and concentration procedures often involving nutrient enrichment, sieving, flotation, centrifugation and filtration. Since most enzyme immunoassays use micro-titre plates or membranes, as the immunosorbent surface, they suffer from slow binding kinetics and small surface areas. In contrast, immunoassays that incorporate magnetic particles (1-10µm) have larger binding kinetics. Antibody coated magnetic beads have the added advantage of utilizing a separation step that is uniquely suited to efficiently extracting antigens from samples that contain particulate debris. Thus, it is possible to develop immunoassays that combine the extraction, concentration and detection of antigens into a single relatively simple and rapid step.

A MAb-based immuno-magnetic bead assay was developed for the detection and quantification of R. solani in soil (Thornton, 1995). Mycelial antigens are solubilized in saline buffer containing detergent and following high speed centrifugation soluble antigen is delineated by the addition either of a mixture of mouse IgA and IgM MAbs from the hybridoma cell lines EH2 and EE1 or a mixture of IgM MAbs and rabbit PAbs. Isolation of the soluble immune complex is achieved by the addition of magnetic beads coated with rat antibodies that specifically recognize the mouse IgM antibodies. Detection of the bound antibody-antigen complex is then accomplished using a commercial antibody-enzyme conjugate and chromogen.

3.2. TRICHODERMA HARZIANUM

3.2.1. PLATE TRAPPED ANTIGEN (PTA)-ELISA's

HD3 and GH5 MAbs were used to develop PTA-ELISA's for the detection and quantification of mycelium and phialoconidia of T. harzianum in peat-bran (Thornton et al., 1993 and Thornton and Dewey, 1995 respectively). Because the two assays incorporate MAbs that each specifically recognize only a single component of fungal biomass it is possible to determine the relative epidemiological roles of mycelial fragments and conidia of Trichoderma spp. in biological control of R. solani. These assays represent a significant improvement on conventional methods involving dilution plate counts and selective media. Unlike the PTA-ELISA's, plate assays do not distinguish satisfactorily between mixed species populations (Trichoderma will rapidly outgrow and occlude colonies of Rhizoctonia, when present at high densities, even on Rhizoctonia-selective media). Plate counts do not distinguish between mycelial fragments and conidia of Trichoderma, yet the two components may have quite different epidemiological significance in the control of *R. solani*. Conidia frequently remain dormant in soils while mycelial fragments germinate, grow, infect a parasite and lyse. Hence the production of spores by a hyperparasite population of *Trichoderma* spp. may swamp ecologically important changes in mycelial fragments.

4. QUANTIFICATION OF IMMUNOLOGICAL ASSAYS

4.1. QUANTAL RESPONSES

Quantal responses yield estimates of the density of colony forming units (cfu's) g^{-1} soil of the pathogen or hyperparasite. Quantal assays are determined by the presence or absence of a detectable reponse in replicates of successively diluted aliquots of test soil. Samples of infested soil are taken and a set of two-fold diluitons prepared with 5-10 replicates per dilution. The replicates are incubated for 48h, at $25^{O}C$, with an aqueous semi-selective medium, to amplify the biological signal (Thornton *et al.*, 1993), after which the absorbance is determined by ELISA. The replicates are scored for the presence or absence of the fungus, relative to the uninfested control samples. Maximum likelihood methods are then used to estimate the most probable number of cfu g^{-1}. Estimates are then compared with cfu g^{-1} obtained by maximum likelihood estimation of the most probable number for direct plate culture methods.

4.2. QUANTIFICATION OF FUNGAL BIOMASS

Using ELISA it is possible to make quick, direct estimates of fungal biomass in soil that do not require the establishment of dilution series. Fungal antigens are solubilized using buffers, particulate debris eliminated using high speed centrifugation and micro-titre wells coated with soil extracts for analysis by ELISA. Quantification of biomass from absorbance values of soil extracts requires the establishment of standard calibration curves. In their simplest form this involves coating wells with spores, for example phialoconidia of *T. harzianum*. Prerequisites of standard calibration curves for mycelial biomass estimations are that they are simple to establish, replicate and repeat and that they do not suffer from extremes in variability. For estimations of mycelial biomass of *R. solani* and *T. harzianum* stock sources of lyophilized mycelium are used. Extracts are diluted in the appropriate buffer to establish a range of absorbance values from which estimates of biomass can be interpolated.

5. EPIDEMIOLOGICAL SYSTEM

The epidemiological system involves the addition of *R. solani* to establish populations in media of increasing microbial complexity comprising sand, partially sterilized field soil and unsterilized field soil, with and without a host and in the presence and absence of *T. harzianum*. The total volume of the system is in the order of 1-2L. This is a simple experimental system that is easy to maintain and monitor but which allows complicated dynamics due to saprophytic and (hyper)parasitic activity of *R. solani* and *T. harzianum*. A cycle takes 30-40 days in the absence of the host, during which the density of *R. solani* rises, following the production of secondary inoculum (P_2) by saprophytic activity of the primary inoculum (P$_1$), and thereafter declines. Addition of *T. harzianum* suppresses the activity and multiplication of *R. solani*. Consistent and reproducible results have been obtained for the analysis of microcosms by ELISA and by conventional plate methods (see *Figure 2.*, with further details in Thornton *et al.*, 1995b).

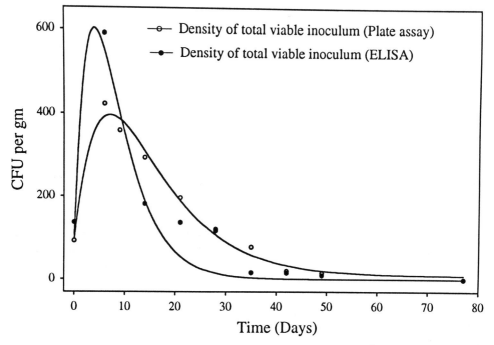

Figure 2. Trajectories showing changes in estimated inoculum densities (cfu g^{-1}) of *R. solani*, in an epidemiological microcosm, using ELISA and conventional dilution plate assay. Independent analysis of changes of P$_1$ and P$_2$ suggests that the ELISA is more sensitive in the detection of high densities of small propagules.

6. CONCLUSIONS

We have shown that it is possible to raise MAbs to live mycelium of *R. solani* (Thornton *et al.*, 1993) that are species-specific and genus-specific MAbs that differentiate between live mycelial fragments and conidia of the hyperparasite *T. harzianum* (Thornton *et al.*, 1993 and Thornton and Dewey, 1995 respectively).

By comparing quantitative MAb-based assays with conventional plate culture methods we have established that immunological assays can be used to detect and quantify viable propagules of these organisms in media of increasing microbiological complexity comprising model reproducible sand-based epidemiological systems to unsterilized soil. We have developed and tested a quantitative assay based on maximum likelihood estimation of experimental epidemiological microcosms to quantify and analyse population cycles of *R. solani* in the presence and absence of *T. harzianum* (Thornton *et al.*, 1995b). We have developed simple mathematical models to describe the observed behaviour of *R. solani* and *Trichoderma* spp. in soil (Thornton *et al.*, 1994b).

We have developed and tested a non-destructive immunoblotting technique in order quantify the temporal and spatial spread of *R. solani* in rhizosphere and bulk soil, *in situ*, in the presence and absence of *T. harzianum* (Thornton *et al.*, 1995a).

7. REFERENCES

Gilligan, C.A. (1983) Modeling of soilborne pathogens, *Ann. Rev. Phytopathology* 21, 45-64.

Gilligan, C.A. (1990) Antagonistic interactions involving plant pathogens: fitting and analysis of models to non-monotonic curves for population and disease dynamics, *New Phytologist* 115, 649-665.

Ogoshi, A. (1987) Ecology and pathogenicity of anastomosis and intraspecific groups of *Rhizoctonia solani* Kuhn, *Ann. Rev. Phytopathology* 25, 125-143.

Thornton, C.R. (1995) Detection and quantification of mycelium of *Rhizoctonia solani* in soil by monoclonal antibody-based immuno-magnetic bead assay, (in press, *Soil Biol. Biochem.*).

Thornton, C.R. and Dewey, F.M. (1995) Rapid detection of phialoconidia of *Trichoderma harzianum* in peat-bran by monoclonal antibody-based enzyme-linked immunosorbent assay, (in press, *Mycological Research*).

Thornton, C.R., Dewey, F.M. & Gilligan, C.A. (1993) Development of monoclonal antibody-based immunological assays for the detection of live propagules of *Rhizoctonia solani* in soil, *Plant Pathology* 42, 763-773.

Thornton, C.R., Dewey, F.M. & Gilligan, C.A. (1994) Development of a monoclonal antibody-based enzyme-linked immunosorbent assay for the detection of live propagules of *Trichoderma harzianum* in a peat-bran medium, *Soil Biol. Biochem.*, 26 (7), 909-920.

Thornton, C.R., Bailey, D.J., Gilligan, C.A. and Dewey, F.M. (1995a) A non-destructive immuno-blotting technique for the temporal and spatial visualization of growth of *Rhizoctonia solani* mycelium '*in situ*'. (in preparation).

Thornton, C.R., Gilligan, C.A., Bailey, D.J. & Dewey, F.M. (1995b). Quantification and modelling of population cycles of *Rhizoctonia solani* with a monoclonal antibody-based immunological assay (in preparation).

8. ACKNOWLEDGEMENTS

The author would like to thank Drs. F.M. Dewey and C.A. Gilligan for supervision during the course of studies performed at Oxford and Cambridge Universities and for technical assistance from Messrs. D.J. Bailey and B.V. Goddard.

Group discussion *in plenum*

Non disruptive *in situ* detection methods.

Presentation of group solution

LARS BOGØ JENSEN
*Copenhagen University, Department of General
Microbiology, "Splejsen"
Sølvgade 83 H
1307 København K
Denmark*

During recent years the development of *in situ* on line detection methods
of single cells in complex microbial communities has extended the scope
of microbial ecology. These new methods have made it possible to monitor
microbial communities without the isolation and characterisation of
individual members. Such studies revealed that the survival strategies of
single microbes depend on complex interactions in the microbial
community. These interactions among sub groups in the microbial
community are monitored by the new techniques with no disruption of the
community's ecological niche. These developments should allow the
creation of new non disrupted microcosm where microbial interactions
could be monitored by use of several classical as well as contemporary
methods. Relying on a combination of methods rather than on a single
would give us a better in depth picture We therefore suggest that the
following microcosm for studies in soil should be designed (see Figure 1)
to answer the questions posed to this group.

To monitor the fate of the introduced antagonist the microorganism must
first be isolated and characterised. Insertion of a marker gene as *gus*
(Liljeroth *et al.* 1993, Roberts, I. N. *et al* 1989), *luc* (Jansson, J. *et al*
1995), *lux* (Prosser, J.I.1994) or *gfp* (green fluorescent protein)(Chalfie,
M. *et al* 1995) into the antagonists genome will enable tracking it in the
microcosm either by selective plating (disruptive technique) or by use of
photon cameras or confocal laser scanning microscopy (non disruptive).
Here hyphe length should be measured. A marker gene could also be

D. F. Jensen et al. (eds.), Monitoring Antagonistic Fungi Deliberately Released into the Environment, 155–158.
© 1996 *Kluwer Academic Publishers. Printed in the Netherlands.*

cloned downstream from a promoter for a gene specifying the antagonistic trait and the activation of this gene can then be monitored *in situ*.

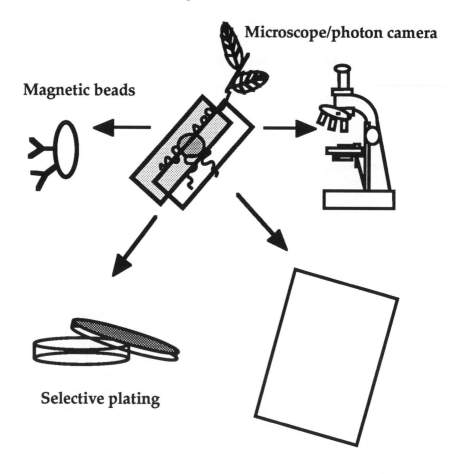

Figure 1

The microcosm is a box made of object glass plates for microscope where non sterile soil is placed in between. A seed, the pathogenic fungi and its antagonist are added to the microcosm. Growth of the seedlings are monitored and the effects of the presence/absence of pathogen and antagonist are determined. The entire microcosm can be placed directly in a microscope.

The overall effect of the antagonist on pathogenic fungi can be monitored by the biomass of the plant that is introduced as a seedling to the microcosm.

The presence of the pathogenic fungi can be monitored by use of rRNA probes (Henrion, B. *et al.* 1992, Li, K.-N., *et al* 1994) and/or monoclonal antibodies (Dewey, F.M. *et al* 1992, Thornton, C.R. *et al* 1994). The use of microscopy will then detect the presence of both the marked antagonist and the pathogenic fungi and their spatial distribution on the seedling and in the surrounding soil. The use of specific antibodies attached to magnetic beads will enable us to extract the fungi from the microcosms without having to use difficult isolation techniques. This technique will also enable us to extract growing spores after addition of nutrients. Using antibodies against proteins essential for infection of the plant pathogen and taking an implant on a nitrocellullose paper of the microcosms will enable us to specifically detect infection points and correlate this to the activity of the antagonist. Vital staining and respiration techniques for the two fungi will tell us how many of the fungi are physiologically active.

Bacteria might enhance the effect of the antagonistic fungi by specific interactions. The presence of bacteria can be monitored using a eubacterial 23S rRNA probe (Amann, R.I. *et al* 1992, Stahl, D.A. and Amann, R.I. 1991).

Using computer software the images of the position of all the microorgansims and their activity will give us a more comprehensive picture of the spatial distribution of the members of microbial community, their interactions, their survival strategies, activities and mutual effects *in situ*. Combining these result with the vital staining and respiration measurements will give a better understanding of the complex three dimensional structure of a microbial population and enable us to conceive ways of improving the ability of the antagonist to survive and perform in soil community.

References:

Amann, R. I, Stromley, J., Devereux, R., Key, R., Stahl, D.A. (1992.) Molecular and microscopic identification of sulphate-reducing bacteria in multispecies biofilms. Appl. Environ. Microbiol. **58**, 614-623.

Chalfie, M. Tu, Y., Euskirchen, G., Ward, W.W., Prasher, D.C. (1994). Green fluorescent protein as a marker for gene expression. Science **263,** 802-805.

Dewey, F.M., Grose, M.J., Twiddy, D.R., Phillips, S.I., Wareing, P.W. (1992). development of a quantitative monoclonal antibody-based immunoassay for *Humicola langinosa* and comparison with conventional assays. Food and Agriculture Immunology, **4**, 153-168.

Henrion, B., Letacon, F., Martin, F. (1992.) Rapid identification of genetic variation of ectomycorrhizal fungi by amplification of ribosomal RNA genes. New Phytologist **122**, 289-298.

158

Jansson, J. K. (1995). Tracking genetically engineered microorganisms in nature. Curr. Opinion in Biotechnology. **6**, 275-283

Li, K.-N., Rouse, D.I., German, T.L. (1994) PCR primers that allow intergeneric differentiation of Ascomycetes and their application to *Verticillium* spp. Appl. Environ. Microbiol., **60**, 4324-4331.

Liljeroth, E., Jansson, H-B., Schäfer, W. (1993). Transformation of *Bipolaris sorokiniana* with GUS gene and use for studying fungal colonization of barley roots. Phytopathology **83**, 1484-1489.

Thornton, C.R., Dewey, F. M., Gilligan, C.A. (1994). Development of a monoclonal antibody-based enzyme-linked immunosorbent assay for the detection of live propagules of *Trichoderma harziamun* in peat-bran medium. Soil. Biol. Biochem. **26** (7), 909-920.

Prosser, J.I. (1994). Molecular marker systems for detection of genetically engineered microorganisms in the environment. Microbiology **140**, 5-17

Roberts, I.N., Oliver,R.P., Punt, P.J., van den Hondel, C.A. (1989). Expression of the *Escherichia coli* beta-glucuronidase gene in industrial and phytopathogenic filamentous fungi. Curr. Genet. **15** (3), 177-180.

Stahl, D.A. and Amann, R.I. (1991). Development and application of nucleic acid probes, p. 205-248. *In* E. Stackebrandt and M. Goodfellow (ed.), Nucleic acid techniques in bacterial systematics. John Wiley & Sons, New York.

TECHNIQUES FOR MONITORING *TRICHODERMA* IN THE PHYLLOSPHERE

Results of a Group Discussion

LINDA HJELJORD
Norwegian Crop Research Institute, Plant Protection
Fellesbygget, N-1432 Ås, Norway

1. Model:

Participants at the Workshop were asked to judge the suitability of the monitoring techniques to a given application. Our group chose to evaluate methods for monitoring the persistence and activity of the antagonistic fungus, *Trichoderma harzianum* P1, in a biological control program in which the antagonist is sprayed on apple blossoms to prevent infection by *Botrytis cinerea*, a pathogen causing dry eye rot in the mature fruit. The model presented is based on a Workshop poster by Hjeljord, Klemsdal and Tronsmo.

2. Need for monitoring

The aim of the monitoring program will be to clarify questions falling into two main categories: risk assessment and elucidation of biological control mechanisms.

2.1 RISK ASSESSMENT

Both ecologists and consumers are expected to have questions about this biological

D. F. Jensen et al. (eds.), Monitoring Antagonistic Fungi Deliberately Released into the Environment, 159–166.
© 1996 *Kluwer Academic Publishers. Printed in the Netherlands.*

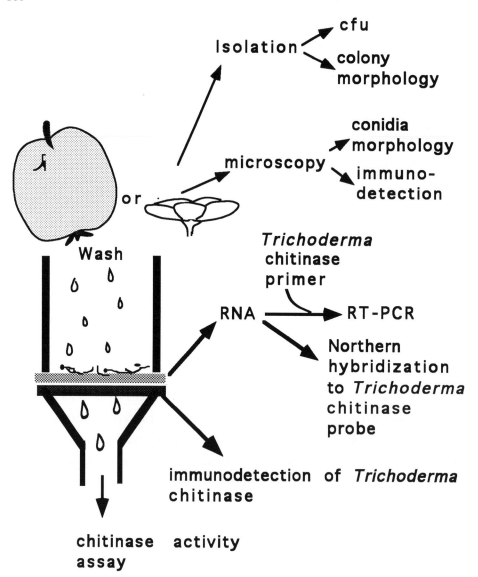

Fig 1. Model for monitoring *Trichoderma* in the phyllosphere

control application. The target area, i.e. the mature fruit, is a food not necessarily washed before consumption. Is there a risk involved in eating apples sprayed during the flowering period with *Trichoderma* ? What about allergy to spores? Introducing a soil fungus such as *T. harzianum* into the phyllosphere will have at least a local effect on the indigenous microflora, and the results of this ecological disturbance are unknown. Will *Botrytis* react to the stress of antagonism by producing increased amounts of toxic metabolites? Will the treatment affect the blossom negatively? Can the introduced isolate be washed off and establish itself in the soil? Some of these questions require special investigation, while some can be answered (or defused) by monitoring the persistence of introduced *Trichoderma* at the site of application.

2.2 ELUCIDATION OF BIOLOGICAL CONTROL MECHANISMS

Antagonist mechanisms need to be clarified with respect both to optimalizing application conditions and formulation and to understanding the organisms´ biological interactions. Does the agent remain at the target site long enough to prevent infection? Is it actively antagonistic? What is the best time to apply the antagonist with regard to biocontrol effect? Answering these questions necessitates monitoring the activity as well as the persistence of the introduced *Trichoderma*.

3. Which monitoring techniques would be most appropriate in our model?

Using a combination of techniques would both increase the monitoring specificity and be most practical, given the need to detect both presence and activity of the antagonist in order to evaluate its effectiveness in biological control. We evaluate the suitability of the various monitoring techniques for use in our model in the following discussion.

3.1 TECHNIQUES FOR DETECTION AND QUANTIFICATION OF THE PRESENCE OF THE ANTAGONIST

The presence and colonization of *Trichoderma* can be monitored using traditional light microscopy, with FDA staining to determine conidia viability, and cfu determination on *Trichoderma*-selective medium. Microscopy is the most simple and direct method of actually confirming colonization by an antagonist or pathogen in a habitat allowing visual distinction of species, for example by conidial or mycelial morphology. Another advantage of using traditional ("old and boring") detection techniques is the public acceptance of such methods compared to those based on genetically modified organisms and the relative ease of instigating such investigations, as no special permission for release of a GMO is necessary. Furthermore, these straightforward techniques would be suitable for monitoring experiments under field conditions, where facilities for sophisticated procedures are often limited. Disadvantages are the limited number of samples possible to monitor by microscopic examination, as well as the time required for isolation and incubation on growth media. Moreover, it is important to keep in mind that cfu counts alone give little information about the status of the antagonist in its actual habitat: fragmented mycelia combined with conidia will produce highly variable numbers of cfu, and increased conidia production can actually be an indication of decreasing vitality of the fungal colony, as Jim Lynch pointed out in his lecture.

Many of the potential monitoring methods discussed during the Workshop are based on molecular biological techniques currently being used to identify fungal isolates: RFLP, RAPD and UP-PCR, isoenzyme analysis and secondary metabolite assays all have in common the need for extensive preliminary work to find suitable probes, primers, enzymes or metabolites that produce patterns which identify and distinguish isolates. Once such sequences have been found, these "fingerprinting" techniques can be sensitive and eventually even routine, but still time-consuming, as they involve isolating and separating the material to be analyzed before the actual assay can be performed.

As explained by Sergey Bulat and Mette Lübeck at the Workshop, universal PCR-primers have been used successfully to differentiate fungal species in the phyllosphere, and they are currently trying to find universal primers able to distinguish *T. harzianum* isolates. Søren Rosendahl showed how isoenzyme bands can be used both as isolate-specific genetic markers and quantitative indicators of activity, and Dan Funck Jensen described the use of isoenzymes to monitor *T. harzianum* isolates in field trials. However, the many steps involved in these methods of identification limit the number of samples that can be screened in a reasonable length of time. Since our model involves monitoring the biocontrol agent in a location not naturally inhabited by *Trichoderma* species, distinguishing the introduced *T. harzianum* from indigenous fungi in the phyllosphere is not anticipated to be a problem great enough to warrant such time-consuming methods. However, for determining the extent to which the agent can establish itself in the soil, e.g. from shed blossom petals, these techniques would provide the necessary isolate-specific identification.

Secondary metabolite profiles are also being used to identify specific isolates, and as such could theoretically be used to identify a specific *Trichoderma* isolate in the phyllosphere. However, as Ulf Thrane pointed out, the low concentrations of chemical "markers" in natural environments, as opposed to in pure cultures in the laboratory, currently limit the use of secondary-metabolite-based techniques in field monitoring applications.

An effective way of circumventing the need for finding isolate-specific "markers" is the introduction of a foreign gene sequence encoding a measurable product, along with an appropriate constitutive promotor, into the genome of the isolate of interest. Techniques for detection of reporter gene products in situ or in extracts have been adapted for use in a variety of habitats. As discussed by Helge Green and Dan Funck Jensen, the GUS (ß-glucuronidase) reporter gene has proven valuable in monitoring *T. harzianum* in the rhizosphere, where X-gluc-stained hyphae distinguish introduced *Trichoderma* from indigenous fungi. As techniques for transforming this isolate with the GUS gene have already been worked out, GUS transformation should be an effective monitoring method for use in our model. The advantage of this technique is the ease of detection of the agent; a disadvantage is the bureaucracy involved before GMOs can be released in the environment on a large-scale basis.

During the plenum discussion of our model, a convincing argument was made for detecting and identifying conidia of both antagonist and pathogen by use of monoclonal antibodies. As explained by Molly Dewey and Chris Thorton, bound antibodies can be detected in situ or in washes by several techniques, and measured through image quantification. Monoclonal antibodies to both *T. harzianum* and *Botrytis cinerea* are currently available, thus eliminating the need for time-consuming preliminary antibody preparation. This method would appear to be the least time-consuming technique for screening samples in our model, and has the advantage of allowing in situ visualization of the presence of both antagonist and pathogen.

3.2 TECHNIQUES FOR MEASURING ACTIVITY OF THE ANTAGONIST

We chose to use extracellular chitinase production as an indication of activity, since this cell-wall degrading enzyme is generally considered important to the antagonism shown by *T. harzianum* against a range of plant pathogenic fungi. A number of tools are available for monitoring chitinase production: the endochitinase gene of *T. harzianum* has been sequenced (by the collaboration of Gary Harman's and Arne Tronsmo's groups in New York and Norway, respectively), allowing design of specific primers and probes. Polyclonal antibodies have been made against *T. harzianum* endochitinase. Furthermore, exo- and endochitinases of *T. harzianum* have been characterized by IEF, as described in Charlotte Thrane's poster at the Workshop. Finally, protocols are available for measuring chitinase activity on a variety of commercially available substrates.

Given the availability of appropriate probes, northern blots of RNA collected under varying times and conditions are commonly used to investigate changes in enzyme production. Northern analysis can be very specific, but in order to accurately monitor low levels of a given transcript, it is necessary to isolate microgram amounts of undegraded total RNA. RT-PCR (based on PCR amplification of cDNAs arising from reverse transcription of total or specific mRNAs) is one of the newer RNA-based methods we considered for detection of chitinase production by *Trichoderma* antagonists. PCR primers can be designed for specific amplification of cDNA encoding endochitinase. Magnetic bead techniques and other "kit" methods allow mRNA isolation from even very small samples of cell lysates, and PCR can in theory detect even single transcripts. However, we felt that the difficulties usually involved in avoiding even partial RNA degradation in the samples would probably make monitoring techniques based on RNA isolation impractical to carry out under field conditions, and unreliable for quantitative investigations.

If specific and sensitive enough, enzyme activity assays in situ or in washes would be more feasible. A fluorogenic chitin substrate would allow in situ visualization and quantification of temporal and spatial variations in the antagonist´s chitinolytic activity, as illustrated in a Workshop poster by Morten Miller. However, enzyme-based assays are limited by the availability of the enzyme of interest, as well as by the need to keep the sample proteins intact and at a representative level of activity. Furthermore, in our model, interfering background activity by plant chitinases and chitinases from other microbes would have to be excluded or taken into account.

One method of checking that the chitinase activity measured in the enzyme assays actually originates from the introduced *T. harzianum* isolate would be to investigate the isoelectric point of the sample enzyme meaured. As mentioned above, chitinases from *T. harzianum* have been found to have isolate-specific isoelectric points. Such analysis would, however, probably require concentrating the sample enzyme .

In our model, it would be easier to perform Western analysis on dot blots of washes from the blossoms, in order to confirm the presence of extracellular chitinases from *T. harzianum*. Even though the *T. harzianum* chitinase antibodies are not isolate-specific, there is but a small likelihood of other isolates of this species being present in the phyllosphere.

The general activity of GUS-transformed isolates can be measured through spectrophotometric assay of extracted ß-glucuronidase, as pointed out by Helge Green and Erland Liljeroth. This activity assay would be useful to correlate with chitinase measurements, to see whether the sample chitinase levels increase specifically, not simply as a reflection of increasing colonization.

4. Conclusions

It is apparent that in our model, monitoring will be improved by a combination of techniques. In the phyllosphere, the sensitivity of the detection methods will be more important than their specificity, as distinction from other *Trichoderma* isolates will probably not be necessary. The presence of the introduced antagonist <u>in situ</u> or in washes can be indirectly detected using monoclonal antibodies or a GUS-transformed antagonist. Colonization by the antagonist and the pathogen can be visualized directly by microscopy, and screened indirectly on selective media by cfu counts and FDA viability stains. If necessary, the isolate identity can be confirmed using DNA fingerprinting techniques or isoenzyme analysis. The activity of the introduced antagonist can be measured by chitinase enzyme assays, the specificity of which can be improved by correlation with general activity assays, e.g. using a GUS transformant, and western analysis using *T. harzianum* chitinase antibodies. If the sensitivity of the enzyme assays proves to be too low for monitoring antagonistic activity, the more difficult RT-PCR procedures may be warranted.

5. Acknowledgements

The following Workshop participants contributed to this report through their active and fruitful discussion of the monitoring techniques presented: Sergey A. Bulat, Dan Funck Jensen, Anne Mette Madsen, Morten Miller, Stina Petersson, Ulf Thrane, Sari S. Timonen, and Valerie Vasseur. Time limits have precluded my submitting this manuscript to the other group members for comment, and thus any errors are my responsibility.

INDEX

168

Developments in Plant Pathology

1. R. Johnson and G.J. Jellis (eds.): *Breeding for Disease Resistance.* 1993
   ISBN 0-7923-1607-X
2. B. Fritig and M. Legrand (eds.): *Mechanisms of Plant Defense Responses.*
   1993   ISBN 0-7923-2154-5
3. C.I. Kado and J.H. Crosa (eds.): *Molecular Mechanisms of Bacterial
   Virulence.* 1994   ISBN 0-7923-1901-X
4. R. Hammerschmidt and J. Kuć (eds.), *Induced Resistance to Disease in
   Plants.* 1995   ISBN 0-7923-3215-6
5. C. Oropeza, F.W. Howard, G. R. Ashburner (eds.): *Lethal Yellowing:
   Research and Practical Aspects.* 1995   ISBN 0-7923-3723-9
6. W. Decraemer: *The Family Trichodoridae: Stubby Root and Virus Vector
   Nematodes.* 1995   ISBN 0-7923-3773-5
7. M. Nicole and V. Gianinazzi-Pearson (eds.): *Histology, Ultrastructure and
   Molecular Cytology of Plant-Microorganism Interaction.* 1996
   ISBN 0-7923-3886-3
8. D.F. Jensen, H.-B. Jansson and A. Tronsmo (eds.): *Monitoring Antagonistic
   Fungi Deliberately Released into the Environment.* 1996
   ISBN 0-7923-4077-9

KLUWER ACADEMIC PUBLISHERS – DORDRECHT / BOSTON / LONDON